Developments in Valve-Regulated Lead-Acid (VRLA) Batteries

DEVELOPMENTS IN VALVE-REGULATED LEAD-ACID (VRLA) BATTERIES

P. G. BALAKRISHNAN AND JOHN RETHINAM

Nova Science Publishers, Inc.
New York

LIBRARY OF CONGRESS CATALOGING-IN-PUBLICATION DATA

Available Upon Request
ISBN: 978-1-60456-269-9

Published by Nova Science Publishers, Inc. ✢ New York

CONTENTS

PREFACE

Although Gaston Planté invented the lead-acid battery in 1859 in France, it took about a century to make it maintenance-free in the sense that there is no necessity for topping up the battery with water. This was made possible by the invention of a hermetically sealed nickel oxyhydroxide-cadmium battery. Though such batteries were variously known in the beginning as a maintenance-free battery, sealed battery, starved electrolyte battery, absorbed electrolyte battery, etc., due to heavy litigation, particularly in the U.K., the term valve-regulated lead-acid battery (VRLAB) was coined to obviate the legal battle. The VRLAB operates on the principle of "internal oxygen cycle" without resorting to the use of catalytic recombination plugs. The binary alloys used in the initial periods of development of VRLAB had no aluminium or silver. After the introduction of Ca-Al master alloy for preparing the Pb-Ca-Al alloy, the grid casting problems were minimized. The introduction of tin to alleviate the problem of premature capacity loss helped to a very great extent in increasing the market share of VRLAB, particularly in the stationary market segment. Developments in the absorptive glass mat (AGM) separator also gave a great impetus to the popularization of the VRLAB. The work done by the Advanced Lead-Acid Battery Consortium (ALABC) paved the way for increasing the Electric Vehicle (EV) market for VRLAB. Rapid charging of VRLAB not only helped in extending the range of EVs by reducing the down time of EVs, but also increased the life of fast-charged VRLAB. The VRLA battery technology being a comparatively new one, a lot of teething troubles were encountered in the initial periods of its introduction. A detailed analysis of the failure modes of VRLAB and remedial measures adopted by the industry is critically analyzed. Extensive trials of VRLAB have been carried out to prove the utility of automotive VRLAB in Australia, Italy, Germany and Japan. Aircraft VRLAB have been proved immune to thermal runaway even though the temperature rises to 72°C. A new idea of a hybrid AGM-Gel VRLAB for use in high temperature operation is under investigation and field trial.

Keywords: VRLAB, manufacture of VRLAB, developments in VRLAB, AGM, alloys, applications, failure modes.

Chapter 1

HISTORY OF THE LEAD-ACID BATTERY

Gaston Planté [1] invented the lead-acid cell in 1859 in France and to perpetuate his memory a stationary type lead-acid battery has been named "Planté battery".

However 60 years before Planté introduced his $Pb/PbO_2/H_2SO_4$ couple, Alessandra Volta [2] in Italy found out in 1796 that when different metallic discs were stacked in an alternating fashion with moist paste board separators saturated with brine being interposed between them, such an arrangement ("Volta pile") could provide significant electrical current continuously. This was the first "primary battery". The foundations for this device were laid by the works of Luigi Galvani [3], a doctor of medicine and a professor of anatomy in the University of Bologna, Italy in 1780's. In 1786, he had observed that if brass hooks were used to suspend dissected frogs from iron fencing, flexing and twitching of the frog's muscle occurred without any other influence, as for example, lightning.

Being a professor of anatomy, he thought that animal tissue possessed "animal electricity". It fell to Volta to explain that the muscular flexing observed by Galvani in his experiment with frog's legs was of metallic nature in origin and he reasoned that the electricity was due to an "electrical fluid" flowing between two dissimilar metals (iron and brass) and a moist conductor (e.g. biological fluid in the animal tissues). The legs were acting as galvanometers.

To perpetrate the names of the scientists who did earliest innovative work, such sources of chemical power were called "galvanic cells" and the driving force of these cells "volt".

Planté's battery consisted of nine cells in parallel; each cell was constructed by winding two lead strips with a rubber strip, heavy lines or felt placed between them. This arrangement was placed inside a cylindrical jar. After filling the cells with dilute sulphuric acid, the plates were "formed" by using Bunsen battery. This is akin to "planté" formation.

Until 1880's Plante's battery could not find wide-spread use. The "formation" was very costly, tedious and time consuming, because the primary batteries were exhausted quickly and also they could deliver small current only With the development of electromechanical dynamo by Siemens in 1866 [4] as a cheap, reliable practical source of DC for battery charging the popularity of Planté battery increased.

A significant milestone in the commercialization of the lead-acid battery was the ingenious and innovative method invented by Camille A Faure in 1881 [5], again French chemist. This method today is known as "pasting" the grid plate, flat plate or pasted plate. Initially such plates were poor in life due to poor adhesion of the paste to the grid, but latter workers improved upon it by designing several types of grids. Notable among them are Sellon [6] who alloyed with antimony and Volkmar, Swan and others. Commercial productions were started in 1880s in several countries like France, the UK, the USA and the USSR. Most of the initial applications were stationary in nature, but by 1900 lighting of houses, factories and trains assumed great importance.

In 1890's batteries with 9 Wh kg^{-1} specific energy were used for electric road vehicle in USA. In 1912, about 100 electric vehicle manufacturers built 6000 electric passenger cars and 4000 commercial cars in the USA [7].

Twenty years after Plante's discovery of lead-acid system, Gladstone and Tribe explained (1880) [8-9] the reactions occurring in the system on the basis of "double sulphate theory", meaning that both PbO_2 and Pb were converted into $PbSO_4$ during discharge. Although this theory was disputed initially, G.W. Vinal who wrote the classic books on batteries [10] undoubtedly established it.

Chapter 2

Development of the Valve-Regulated Lead-Acid Battery

Scientists and engineers throughout the world have been contemplating ways to reduce the cumbersome maintenance procedure associated with the use of conventional lead-acid batteries. The latter require periodic addition of approved water to make up for the loss of water linked with the side reaction of electrolysis of water during the recharge process. Also, acid-fume laden gases emanate from the battery while charging is going on. This fume will affect the surrounding equipment as well as the atmosphere. Only in the late 1960's, the real "maintenance-free" batteries were realized commercially. The advantages of a VRLA battery are numerous and they are listed below.

Merits or Advantages of VRLA Battery Technology

1. The greatest advantage of VRLAB technology is that such batteries do not require water addition through out their life. As the VRLAB is fitted with a one-way release valve, there is no fear of acid spillage. Thus it is a 'clean' battery.
2. The VRLAB is a "user-friendly" battery in the sense that it evolves negligible amount of gasses in the normal course of operation and there is no acid spray. Hence they can be installed in the office/domestic conditions and in small equipment they can be integrated into the cabinet itself. Thus corrosion of the equipment or

metal surface within the place where batteries are installed is insignificant.

3. Since the acid is mostly absorbed by the separator or gelled, these batteries can be transported easily and in a simple manner. They are non-hazardous cargo both by air and sea.
4. Due to the same reason as given in 3 above, these batteries can be operated on their sides in racks thus reducing the floor space required. Such an arrangement also facilitates routine check of the terminal voltages.
5. For the same reason as in 2 above, there are no special requirements for VRB enclosures. In some countries, only 25% ventilation required for flooded batteries is specified. This is possible because even in the initial stages, the gasses evolved will be only 10% or less compared to the conventional one and as the cycling proceeds, this figure drops to a very low value.
6. Compared with flooded batteries, the densification of acid in the bottom of gel cells is very negligible and in the AGM batteries it is not as serious as in flooded ones [11]. For example, if 1.230 specific gravity sulphuric acid is used, the following data apply for 6th cycle afterwards in 50 cm tall cells. Because of this, the non-uniform utilization of active materials and corrosion of positive grids are eliminated or reduced, thus prolonging the life of the batteries.
7. Due to the high compaction of the active block with separator and plates no shedding is envisaged in AGM cells and since the PbO_2 particles are prevented from sedimentation in gel cells, the life of such batteries is far better than flooded construction.
8. Because of the purity of the raw materials used and the AGM, the self-discharge rate of VRLA batteries is very low. It is less than 0.1% per day as compared with 0.7 – 1.0% per day for flooded cells.
9. Because of the stratification of electrolyte in flooded cells, more overcharge is required for full charge and equalization of the battery units. However, the percentage of overcharge required is less in the VRLAB.
10. Due to the negligible maintenance required for VRLAB (i.e. no topping up, no cleaning of the corroded terminals, less equalizing charges), the operational cost over the entire life of the VRLAB is very low, bringing the cost of the VRLAB to a level equal to that of flooded counterparts.

11. In certain applications like automotive batteries, the batteries can be installed in any place other than the vicinity of engines and this flexibility of placement of the battery has been appreciated in actual practice by the designers.
12. There is no explosion hazard as the gassing is very limited. Even so, most of the VRLABs have been provided with explosion proof vents for protection against explosion in the event of user abuse.
13. Since VRLABs are factory-charged, it is very easy for installation avoiding the tedium of initial filling and initial charging, thus minimizing the time required for installation.
14. In the event of a severe damage to the container, the acid does not spill out and the added advantage is that even after being shot by bullets, combat vehicle batteries have been found to be still functional giving sufficient power to the equipment.

Table 1. Stratification of electrolyte in different lead-acid cells (50 cm tall cells using 1.230 specific gravity sulphuric acid)

	Battery type		
	Gelled	**AGM**	**Flooded**
Sp. gr. at the bottom	1.240	1.260	1.280
Sp. gr. at the Top	1.232	1.200	1.170

With a view to do away with the periodic addition of water, T.A. Edison in the year 1942 proposed a heated platinum wire or sponge inside the battery to effect a recombination of oxygen and hydrogen evolved in a battery [12]. Several researchers [13-16] concentrated on catalytic recombination of gases. But these two efforts proved commercially unsuccessful. Later emerged the concept of using an auxiliary electrode [17-20] for the above purpose.

A breakthrough in the development of commercially viable sealed LAB came after the invention of sealed nickel cadmium cells. The principle of "oxygen cycle" (indirect recombination of oxygen) was first outlined by Lange[21] in his US patent in 1938. Later Levin and Thompson [22] and Jeannin [23] and a host of other patents [24-25] described the construction of sealed nickel-cadmium cells. Actually Jeannin [26] has as early as 1950 clearly defined the principles to be applied in sealed LAB. Ruetschi and Ockerman [19] described the recombination process which could be expected to occur in a sealed LAB. Devitt and McClelland [27] described for the first time a commercial sealed LAB based on the oxygen cycle principle. The later 1950's and early 1960's paved the strong foundations necessary for the commercial developments of sealed batteries. Simultaneously two technologies, one based on gelled electrolyte (GE) and the other on AGM were developed, the former Germany and the later in USA, Japan and Europe. The process of finer refinements in the technology is a continuing one till date making subtler improvements in several fields of VR battery evolution as well as in application areas [28-49].

To start with, the valve-regulated lead-acid batteries were called "maintenance-free" batteries, sealed batteries and so on. Due to a lot of litigation between the consumers and the manufacturers regarding the use of the term "maintenance-free", the presently used term "valve-regulated" became widely accepted. Since the VR battery has one-way pressure release valves, the usage of the term "sealed" also is discouraged.

The lead-acid battery science and technology has been reviewed by several authors from 1902 to 2002.[50-61]

The science and technology of the valve-regulated lead-acid battery has been reviewed by Berndt in 1993 and 1997 [62], by Salkind and others in 2002 [63] and very recently by Rand et. al. in 2004 [64].

Chapter 3

CRITERIA FOR THE SUCCESSFUL OPERATION OF "OXYGEN CYCLE" IN VRLAB

The basic principles involved are embodied in the "oxygen cycle" (Figure 1) and have been described by various authors [65-74] Basically the oxygen evolved during charging of a VR battery is made to pass through unsaturated pores available in the gelled electrolyte structure or AGM separator and reach the negative plate where it combines with active lead to form PbO, which gets converted to $PbSO_4$. The latter is converted to Pb and H_2SO_4 by an electrochemical route by reacting with the hydrogen ions resulting from the decomposition of water at the positive plates when they are charged.

The reactions are as follows:

At the positive plate:

$$2H_2O \rightarrow 4H^+ + O_2 \uparrow + 4e^- \quad (1)$$

At the negative plate:

$$2Pb + O_2 + 2H_2SO_4 \rightarrow 2PbSO_4 + 2\ H_2O + \text{Heat} \quad (2)$$

$$2PbSO_4 + 4H^+ + 4e^- \rightarrow 2Pb + 2H_2SO_4 \quad (3)$$

The last reaction restores the chemical balance of the cell. The net sum of the reactions (1) to (3) being zero, the electrical energy spent during charge is converted into heat rather than to chemical energy [75].

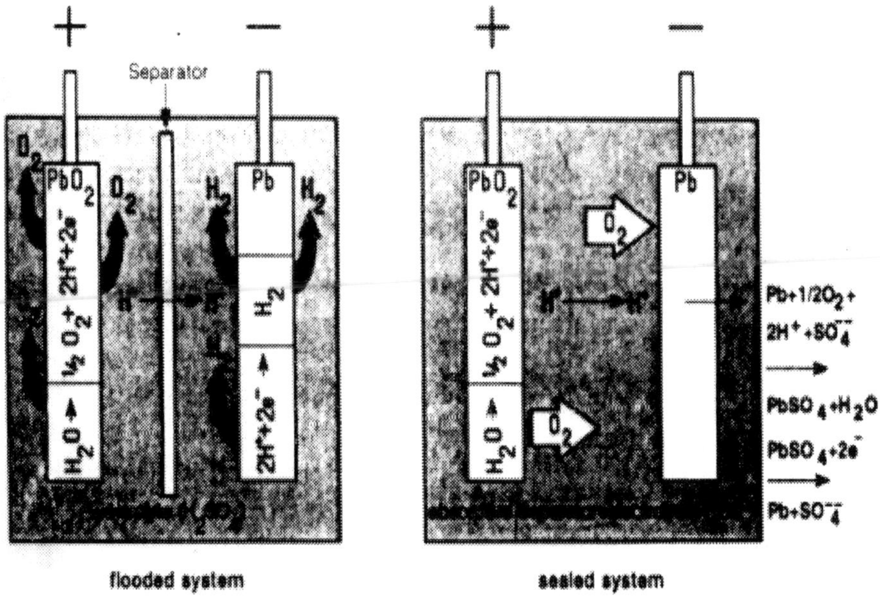

Figure 1. O2 Cycle. Reference: 70 E. Nann.

As a consequence of the chemical reaction (2), the potential of the negative plate becomes more positive ("depolarization" occurs) thus reducing the rate of hydrogen evolution.

A corresponding hydrogen cycle does not exist because the oxidation of hydrogen at the positive plate is too low as a result of the presence of the protective corrosion layer of PbO_2. Thermodynamically, hydrogen oxidation (the reversal of hydrogen evolution) is a possibility above 0 volt vs. standard hydrogen electrode. But it is so much hindered that it can be neglected [75, 76].

For the successful operation of the oxygen cycle in VR battery, the following design aspects are to be considered:

(i) The electrolyte volume is reduced and the same is kept absorbed in porous AGM separator or matrix or gel using fumed silica. To compensate for the decreased volume of electrolyte and to increase the low rate capacity the sp. Gr. is slightly increased.

(ii) The VR batteries have, historically, been designed with excess negative active materials (NAM). However continued developments of such cells have resulted in a reduction in the amount of NAM. In fact present designs generally have excess positive active material (PAM) to improve cycling life capabilities without impairing the recombination efficiency (RE) [71, 77]. An alloy with fairly high hydrogen over voltage should be employed. The alloy may take the form of a low-antimony-cadmium alloy or lead –calcium alloy. A related aspect is the purity of the materials used. Because of the deleterious effects of contaminants on the performance of VR batteries, it is obligatory to use very pure materials.

(iii) A one-way pressure release valve is to be provided. The purpose of such a valve is three fold:

(a) To prevent accidental ingress of unwanted air from the atmosphere this results in the discharge of NAM.
(b) For the effective pressure – assisted transport of the oxygen from PAM to NAM and
(c) To protect the battery from unexpected explosion this may be caused by an abusive charge.

(iv) The plate-separator-container wall compression is an integral part of the design since this ensures not only a good electrolyte diffusion between plate and separator but also enhances life expectancy by inhibiting positive active material expansion and capacity loss For valve regulated lead acid (VRLA) applications involving deep cycling, significant cycle life improvement can be achieved by applying pressure to the plates via the compressible recombinant battery separator mat. This will require changes not only to the separator, but also to the battery design and assembly techniques. Techniques of applying very high pressures to the cell group during battery assembly need to be developed. Changes to the container design and materials may also be needed, and the overall battery design may need to include an internal or external means to maintain the applied pressure throughout battery life. Less compressible separators may also enable higher plate-group pressures to be maintained [78].

(v) To withstand the excess pressure encountered in such batteries, the containers also should be rigid [33, 38]. The container manufacturers should develop new case designs that could assist in maintaining the element pressure throughout the life of VR batteries.

(vi) Since there is no sludging of active materials, the provision of mud space is eliminated.

(vii) An integral seal of cover to container and post to cover should also be ensured [79, 80].

(viii) As a result of the high pressure required to obtain better life from VR batteries, there is a possibility of an adverse effect on the NAM performance. This should be resolved by designing the horizontal and vertical negative grid members to be a little thinner than the grid frame so that the excessive pressure is borne by the frame and not by the NAM [76].

The effects of different factors like the state of charge (SOC), the temperature, the polarization potential, the H_2SO_4 concentration, and the saturation level of the negative plate on oxygen recombination in valve-regulated lead-acid batteries have been investigated by Garche and his co-workers [81]. They have concluded that the increase in the temperature of the negative plate, increase in the negative polarization and also the decrease of the saturation level of the negative plate can all promote oxygen recombination, but the SOC has negligible effect on the oxygen recombination as long as the capacity of the negative plate is high enough. Moreover, the oxygen recombination rate increases and then drops when the H_2SO_4 concentration at the negative plate increases from 0 to 5 M. The experiments also proved that the saturation level of the negative plate is the dominant factor affecting the oxygen recombination rate.

Chapter 4

GRIDS AND ALLOYS FOR VR BATTERIES

Almost all the manufacturers of valve-regulated lead-acid battery manufacturers throughout the globe employ non-antimonial alloys, especially Pb-Ca-Sn alloys with minor silver addition, except a few who use pure lead, Pb-Sn or Pb-Sn-Cd alloys. Calcium is not precipitated on the negative plate, as does the antimony from an antimonial alloy and remains as Ca^{++} ions in the electrolyte [82]. As a consequence, the hydrogen evolution rate is low and remains so throughout the life of the battery. However, the grid is liable to be severely attacked, depending on the composition of the alloy. Antimony is rarely included in the grid alloy material for VR batteries because of its tendency to reduce the hydrogen overvoltage, thus encouraging gassing at the negative plate. Careful screening must be done in order to avoid other similar elements which might reduce the hydrogen overvoltage.

However, there are a few disadvantages in using Pb-Ca alloy. The calcium content has to be kept within 0.09% for good creep resistance, but care should be taken to see that the Ca content does not go below 0.07% for better handling properties. The second disadvantage of the alloy is promotion of premature capacity loss which can be counteracted by incorporating orhto phosphoric acid in the electrolyte and tin in the grid alloy. The third problem is concerned with the deviation of Ca content while casting is in progress. This problem is obviated by inclusion of 0.015-0.03% aluminium in the Pb-Ca alloy and maintaining the molten alloy temperature above 525 C so that the aluminium is in solution.

Lead even when alloyed with Ca is notorious for its poor creep resistance. This aspect has been improved by the addition of 1.0 to 1.5% tin to the ternary Pb-Ca-Al alloy. Detailed studies of such alloys [83-88] reveal that the addition brings additional unexpected benefit of enhanced corrosion resistance of the alloy toward anodic oxidation and reduced electronic resistance of the corrosion layer.

The influence of two levels of calcium and tin, as well as the grid thickness, on the corrosion rate (i.e., growth) was investigated [89] through tests on bare, pasted grids as well as life testing of batteries. A combination of 0.07 wt.% Ca and 0.7 wt.% tin in the alloy, together with grids with 4 mm thickness yields low growth-rates that have sufficient for long-life VRLA batteries (3-6% growth in 54 weeks at 60°C). The microstructure of the alloy, i.e., the number of grain boundaries per unit volume determines the magnitude of the resulting corrosion attack. Lead-calcium-tin-aluminium alloys with an approximate composition of 0.07 wt.% Ca and 0.7 wt.% Sn have a lower number of grain boundaries when compared with alloys of the same system with 0.09 wt.% Ca and 0.3 wt.% Sn. Superior life of such batteries more than offset the penetrating grain-boundary attack observed with the Pb-0.07wt.%Ca-0.7wt.%Sn alloy, as compared with the Pb-0.09wt.%Ca-0.3wt.%Sn alloy.

The mechanical properties, grain structure, and corrosion behaviour of cast and wrought Pb-Ca-Sn and Pb-Sb alloys for tubular grids have been analyzed in detail [90]. The author also suggests manufacturing techniques for high performance, wrought, tubular plates. Alloys with 1-6 wt.% antimony have been used for reduced maintenance. Although these alloys permit the manufacture of such batteries, cast Pb-Ca-Sn alloys are significantly inferior to cast Pb-Sb alloys in mechanical properties. Wrought Pb-Ca-Sn alloys, when used for tubular grids, permit the application of maintenance-free alloys with mechanical properties comparable with, or higher than, those of high-antimony alloys. Wrought materials increase life due to the absence of casting defects. Wrought lead-calcium alloys also offer a dramatic improvement in creep and corrosion resistance compared with conventional cast, tubular, Pb-Ca-Sn alloys, as well as superior conductivity to cast Pb-Sb. Wrought Pb-Ca-Sn alloys permit the production of tubular grids at high speed in shapes and forms that are difficult to produce from cast materials. These grid shapes can lead to higher performance, higher discharge-rate, tubular plates.

Japanese scientists have examined [91] the passivation phenomena of lead-calcium-tin alloy expanded grid and found an optimum composition to avoid the passivation observed when a VRLA battery is deeply discharged. The presence of tin was not to prevent corrosion nor to protect diffusion of oxygen. For the mechanical characteristics tin content in the wrought base alloy was limited to 0.7 wt%. If a higher-tin alloy is required, the author recommends a new processing condition and/or means to be investigated for handling the alloy.

The intergranular corrosion of the positive grid made of Pb-Ca-Sn alloy during float charging is one of the most serious problems [92]. To prevent the corrosion, the electroconductive perovskite type oxide films containing lead metal were formed on the surface of the cathode grid. As the results of surveying various systems between β-PbO_2 and metal oxides or hydroxides, it was confirmed that $BaPbO_3$ and $SrPbO_3$ were formed even at 300°C. The life of trickle-charged battery with the positive grid covered with $SrPbO_3$, was twice longer than that of the standard battery without the perovskite oxide treatment.

Torcheux et al. [93] have studied the influence of grid technology such as gravity cast technology and rolled expanded technology with and without silver addition to the grid alloy material for VRLA-AGM batteries meant for electric vehicle application. Self-discharge tests, deep discharge tests and TC69 cycling were carried out. Gravity cast grids gave a better electrical behaviour. The batteries using rolled expanded technology gave better results in specific energy and better stability during the TC69 test.

Taguchi et al. [94] studied high-temperature corrosion behavior of Pb-Ca-Sn alloys during charge/immersion cycling in 4.5 molar sulfuric acid solutions at a temperature of 348K for a period of 168 hours. The alloys studied were Pb-0.08Ca-Sn alloys with Sn content ranging from 0 to 0.2%. The grid was attacked depending on the composition of the alloy. The result suggested that the corrosion attack occurs mainly at the grain boundaries of the alloy microstructure and 97 to 99% of the charge current is used for the oxygen evolution. Also, 0.08 to 2.75% of the current is spent in formation of the corrosion layer consisting of $PbSO_4$ and ß –PbO_2 and the ratio decreases as the Sn content increases.

Ball and others [95] investigated variations in positive active material composition and structure by forming electrodes from grey oxide pastes containing additions of red lead, tetrabasic lead sulphate, or sulphuric acid (sulphated). Two grid types were employed having either pure lead or a lead/tin alloy. A number of designs of battery were manufactured with different grids and positive active materials. Results indicated that both grid alloy composition and PAM type affect the corrosion layer properties. Ultra-microtoming was used to prepare sections of the grid/corrosion layer interface. Results showed that corrosion propagated along tin rich grain boundaries. These authors have also identified a number of different defect types present in the corrosion layer [96] by examining the failure of a series of 40 Ah valve-regulated lead-acid batteries which have been cycled extensively.

VR batteries subjected to life cycle testing showed a large variation in the number of cycles to failure ranging from 10 to 133 cycles [97]. Results obtained by from destructive examination of these batteries provided information on the probable causes of failure by monitoring reduction in cell voltage during a final discharge/charge cycle, scanning electron microscopy (SEM), BET surface area analysis, X-ray diffraction, interfacial analysis and electron probe analysis. Batteries that failed after a low number of cycles, 10 and 28, were believed to have done so due to sulphation of the positive plate. Such batteries would have been revived easily had they been charged properly. Thick corrosion layers were shown to be the cause of failure in the batteries that sustained high numbers of cycles, 92 and 133. Results suggested that batteries failing at intermediate numbers of cycles, 42, 49, 65 and 73, failed due to degradation of the cells simultaneously and a single failure mechanism could not be identified.

A light weight lead-coated glass fibre mesh grid has been described for use in valve-regulated lead-acid (VRLA) batteries [98]. Plates made with these new grids show a higher material utilization over a wide range of discharge rates at a current density of 20 to 200 mA g^{-1} and in the temperature range of -15 - 25°C. The results suggest that the lead-coated glass fibre grid can replace the conventional gravity-cast grid without causing any deleterious effects so far as the cycle life is concerned.

The influence of various parameters involved in the cast-on-strap (COS) process on the quality of the resultant lug-strap joints has been reviewed by Australian workers [99]. Macroscopic and microscopic features of COS joints in commercially available lead-acid batteries revealed the most common defects that affect the performance and life of batteries. The nature of the fusion at the lug-strap interface for the various lugs and strap alloy compositions used in conventional, hybrid and valve-regulated lead-acid batteries are also discussed.

Chapter 5

CASTING OF GRIDS

Grid casting is the first unit operation in the manufacture of lead-acid battery. Figure 2 is a flow chart giving the various steps involved in the manufacture of valve-regulated lead-acid batteries.

Most of the alloys used are either low-antimony based or calcium-based and hence they have higher freezing temperatures which necessitates keeping the molten metal temperature at higher level compared with conventional alloys. The mould temperature should be around 170-190°C (upper portion) and 210°C for lower portion. The pot and ladle temperatures for manual casting should be in the range of 510-525°C. However in the case of casting machine the pot temperature is usually kept around 400-425°C, but the ladle and pore temperatures are around 510°C.

The addition 0.03% of aluminium to the lead-calcium alloy has served to stabilize the molten metal [100-106] by forming a tenacious protective layer on the molten surface and hence these fears nowadays do not have a base. Because of this, calcium does not dross out and hence there is no necessity of sweetening with calcium-rich lead.

There are several methods to alloy Ca with lead. They are:

i. Direct pyrometallurgical preparation of Pb-Ca master alloy by adding calcium carbide to molten lead in a reducing atmosphere provided by aluminum and subsequent dilution to the required level by adding the master alloy of Pb-Ca-Al thus obtained. Here a flux of CaF_2-$CaCl_2$ is employed to protect the molten lead from atmospheric oxidation.[107]
ii. Using a special sealed dissolution vessel. [105]

iii. Electrochemical route by electrolyzing a mixture of CaF_2-$CaCl_2$ between molten lead electrodes and skimming off the Pb-Ca alloy periodically [108-109].
iv. Mixing the Ca-Al master alloys available commercially with the molten lead [110-114]. Suitable master alloys are CAL-GRID [112-113], Cal-Al™ [114]

Of these the last mentioned method is favoured by most of the manufacturers of the alloy.

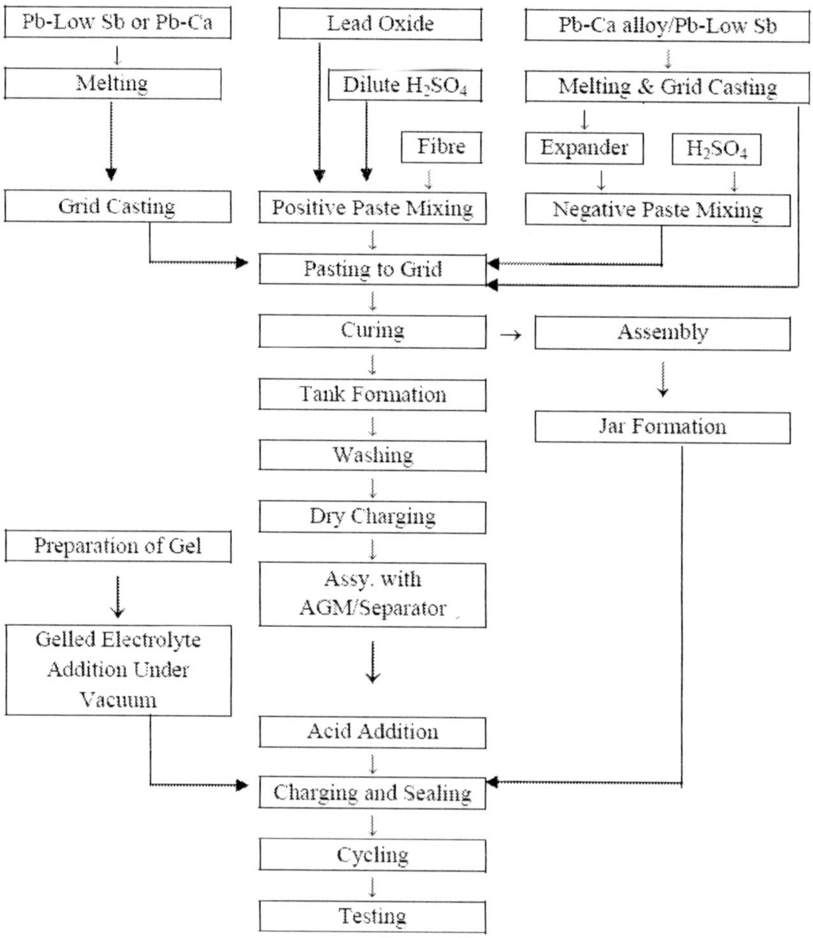

Figure 2. Flow-chart for the manufacture of VRLA Batteries.

Some minor elements like aluminium, silver, bismuth and some alkaline earth metals are also added to lead-calcium alloys to improve the alloy properties and the battery performance. A convenient method for the addition of aluminium into the alloy, compared with other methods reported by Zhong [115]. On addition to lead-calcium alloys, bismuth can form a Bi-Ca compound according to the de-bismuthizing equation: $3Pb_3Ca + 2Bi \rightarrow Bi_2Ca_3 + 9Pb$. There is a slight solubility of Bi_2Ca_3 in the lead at its melting point (327 °C) with an amount of 0.048 wt.% resident bismuth. It is confirmed that the creep strength and the corrosion resistance is distinctly increased by the addition of silver into the grid alloy, but the associated gassing problem should be considered when such alloys are employed in VRLA batteries. Addition of sodium can refine grain sizes and increase the mechanical properties of the grid alloy, but the corrosion resistance of sodium-contained grids is inferior. A general description of experiments and application of tin, aluminium, bismuth, silver and sodium as additives in lead-calcium alloys is given.

The most important precaution is to see that no antimony alloy gets mixed up with lead-calcium alloy and that the rejects as well as scrap should be distinctly marked. Two separate casting sections is a better proposition.

As a consequence of the narrow solidification temperature range of lead-calcium alloys, the alloy solidifies very quickly and this results in the reduced cork-spray life on the moulds. A maximum life of 3 to 4 hours is obtained. Another consequence is that the freshly cast grids are highly flexible and soft necessitating very careful handling in the manual casting operation. In the case of grid casting machine, horizontally mounted gate trim dies are used in the newer generation machine to reduce the bending of grids in the casting machine.

Some manufacturers prefer to give a thin layer of Pb-Sn alloy over the grids [116]. Here the percentage of tin will be much more than in the base alloy. Because of paucity of electrolyte in the VR batteries, a passivating lead sulphate layer sometimes forms on the lead grids during stands, particularly after a deep discharge. Tin is used to combat this passivation phenomenon. A similar function is performed by the electrolyte additive, ortho-phosphoric acid.

Ball [117] arrived at the following conclusions from the investigations carried out on the design of positive grids: Grid bar and positive active material geometry affect corrosion layer thickness. Rotation of grid bars does not result in an increased current density distribution on the bar surface. Current density distribution is greatly affected by grid shape. Due to uneven distribution and thickness of positive active material around the grid bar, a design with sharp corners such as a square section, does not necessarily result in the maximum current density distribution occurring at the corner of the square. Changing the geometry of the positive active material surrounding the grid bar can have a significant effect on reducing the current density distribution.

Grids can also be made from sheet cast lead-calcium alloy or rolled alloy sheets by rolling/punching/expanding. Prengaman [118] suggests that the wrought design offers an improved grid/active material surface area compared with conventional tubular grids. Since the mechanical properties of the wrought material are higher in the longitudinal direction than in the transverse a wrought grid offers an improved resistance to growth. Moreover, such tubular grids are fully dense with hole-free structure that is seen in conventionally cast tubular grids. These grids have no cracks, incomplete fill, or brittle spines despite the mechanical properties of the wrought material. An added advantage is that if a spine becomes distorted, it can be bent back to desired configuration without damage. Wrought tubular grids for VRLA batteries produced from high strength, high tin content Pb-Ca-Sn alloys should be able to be utilized in both gel and AGM constructions.

Analysis of calcium content requires sophisticated equipments like Atomic Absorption Spectrophotometer (AAS). For the shop floor testing of calcium content the "antimony titration" (metallurgical titration procedure) can be adopted. This method was first described by Bouton and Phipps [119]. It was later refined by St. Joe mineral Corporation of the USA into a "ternary bloom test" [120]. For easy accessibility the reader is referred to Hehner's Book [121].

A wet analytical procedure for the estimation of calcium [46] based on calcium oxalate and oxidation of the latter after dissolving in a hot dilute sulphuric acid solution by $KMnO_4$ is given by Clarke and Wooten in 1933 [122].

The lead-calcium-tin grid responds fairly well to age-hardening in a period of 3 to 6 days. [121].

Development of a VRLA battery with thin tubular positive plates having improved specific energy and optimized low charge-factor is described by Dyson and Griffin [123]

The thicknesses of the component is 50% of those for existing technology.This results in an acceptable metal/active-material surface ratio ('gamma' value) by using an elliptical spine. Matching recharge characteristics with cell design and controlled initial electrolyte of the cells is discussed in terms of the influence of the oxygen-recombination rate on the recharge response. Changes during stable operation are illustrated in terms of charge response data during long cycle-life with 80% depth-of-discharge and a low, overcharge, viz., 4% C_5.

Chapter 6

PASTE MIXING AND PASTING OPERATIONS

Paste preparation is the most important unit operation in the manufacture of lead-acid batteries. The paste is the basic material upon which the performance of lead acid battery is built. No amount of skillful manipulations in design or construction can completely correct the deficiency caused by a poorly prepared or incorrectly formulated paste. To prepare a formable paste from lead oxide, a pastable mass is made from lead oxide, water and sulphuric acid. The paste is applied to the grid and cured. In the dry state the pasted plates should have a certain dry strength so that it cannot be rejected in the subsequent manufacturing process.

Table 2. Paste formulae for positive and negative plates [124]

Component	Positive paste	Negative paste
Leady oxide (kg)	3	3
Fibre (g)	0.9	1.8
CMC[@](g) Carboxymethyl cellulose	7.5	-
Stearic acid(g)	-	1.8
$BaSO_4$ (g)	-	11.1
Vanisperse(g)	-	11.1
Carbon black(g)	-	6.3
1.400 sp.gr.H_2SO_4(cm^{-3})	200	200
Water(cm^{-3})	390	330
Paste density(g cm^{-3})	4.5-4.6	4.7-4.8

The normal procedure of paste preparation can be followed with slight modifications to lower the wet paste density for compensating the lesser volume of electrolyte used in finished batteries so that the required low rate capacity can be obtained. The plate thickness, the weight of wet pasted plates and parameters which affect the porosity should be closely controlled to obtain a highly reproducible product.

Lam [124] gives a guidance in the formulation of paste for VRLA battery.

The influence of pastes from various $H_2SO_4/H_2O/PbO$ ratios on the material structure and electrochemical characteristics of PbO_2 electrodes is studied [125]. The structural and electrochemical characteristics of these electrodes were investigated by scanning electron microscopy (SEM), X-ray diffraction, mercury intrusion proximity (MIP), ac impedance, and charge-discharge curves. Three pastes (designated A, B, and Q were examined to assess the influence of $H_2SO_4/H_2O/PbO$ ratios. Paste A, prepared with higher amount of water, had a larger specific surface area and pore size. The group A plates had a greater porosity and exhibited higher active material utilization and higher initial capacity but poor mechanical strength resulting in a higher capacity loss for each cycle. In studies on ac impedance, the appearance of a high-frequency inductive loop in the group A plates indicated that these plates had active materials of greater porosity during the electrode processing reactions. The group C plates, prepared with more H_2SO_4, contained more α-PbO_2 crystals which resulting in a smaller charge-transfer resistance, and a larger exchange current.

A new cureless paste formulation technique for positive plates for valve-regulated lead-acid batteries has been developed [126] which employs hydrogen peroxide to replace the traditionally used sulfuric acid. After pasting, the plates can be assembled into groups and placed directly into a battery case. Formation can then be performed without the plates undergoing the conventional curing process. Such a non-curing process could reduce production time and costs. More significantly, the resulting plates provide good flexibility which is particularly suitable for spirally-wound battery designs.

Nanocrystalline lead oxide was prepared through two-step chemical reactions and was tested as the starting material for electrode active material for the valve-regulated lead acid (VRLA) battery [127]. By using this nanocrystalline lead oxide, the pasted electrodes could be directly formed without a curing process. The initial specific capacity of the new electrode was 30% higher than that of the electrode with the conventional ball-milled leady oxide and the cycle behaviour of the new electrode was also improved.

The effects of adding carbon fibre to red lead positive paste was investigated by Ball [128]. The batteries with the additive were cycled and subjected to destructive examination and analysis using scanning electron microscopy and X-ray diffraction. The major failure mode in a cycled VRLA battery is degradation of the positive active material. Results obtained suggested that the addition of carbon fibre to positive paste increased the capacity by approximately 20% and substantial improvements in cycle life were also observed.

Rand and his co-workers in CSIRO, Australia, have carried out extensive studies on the effect of bismuth in the refined lead. To meet the increasing demand for valve-regulated lead-acid (VRLA) batteries, Pasminco Metals has made a new soft lead. Incorporating bismuth up to 0.05-0.06% in the lead produced a significant improvement in battery cycle life and at the same time problem metals like arsenic, cobalt, chromium, nickel, antimony and tellurium that are known to be harmful to VRLA batteries are controlled to very low levels. Bismuth-bearing oxide has been characterized in terms of phase composition, pal-tide size distribution, BET surface area, and reactivity. The performance of pasted plates prepared from the Bi-bearing oxide and Bi-free oxide virtually no bismuth (< 0.005 wt.%) was examined. Increasing the level of bismuth and lowering those of the other impurities in soft lead produced no unusual changes in either the physical or the chemical properties of the resulting Bi-bearing oxide compared with Bi-free oxide and so there is no need for battery manufacturers to change their paste formulae and paste-mixing procedures on switching to the new Bi-bearing oxide. There is little difference in the rates of oxygen and hydrogen evolution on pasted electrodes prepared from Bi-bearing or Bi-free oxides. On the other hand, these rates increase on the former electrodes when the levels of all other impurities are made to exceed (by deliberately adding the impurities as oxide powders) the corresponding, specified values for the Bi-bearing oxide. The latter behaviour is particularly noticeable for hydrogen evolution, which is enhanced even further when a negative electrode prepared from Bi-bearing oxide is contaminated through the deposition of impurities added to the sulfuric acid solution. The effects of impurities in the positive and negative plates on the performance of both flooded-electrolyte and VRLA batteries are assessed in terms of water loss, charge efficiency, grid corrosion, and self-discharge. The causes of negative-plate discharge in VRLA batteries under float conditions are also analyzed [124].

The use of Bi-bearing oxide yielded higher initial capacity, a significantly faster rise in capacity to the full value with subsequent cycling, a lower rate of self-discharge rate and good cycle-life under the International Electrotechnical Commission (IEC) test [129].

Batteries made from other factory oxide exhibit cycle-lives which exceed the minimum life specified by the Japanese Industrial Standard (JIS) test and by the International Electrotechnical Commission (IEC) test. Under the JIS procedure, the failure of both types of battery is due to expansion of the positive-plate material and subsequent extensive loss in connectivity between the constituent agglomerates of lead dioxide. On the other hand, IEC procedure induces failure by undercharging of the negative plates. The use of VRLA Refined(TM) oxide produces a marked improvement in battery cycle-life. Gains in performance from 675 to 800 cycles and from 510 to 675 cycles are obtained under JIS and IEC tests, respectively. It is suggested that the presence of bismuth at the specified level extends battery life by strengthening and increasing the connectivity in the positive-plate material (JIS test) or by raising the chargeability of negative plates (IEC test) [130].

The mechanism by which bismuth enhances the capacity of valve-regulated lead-acid (VRLA) batteries has been elucidated using model experiments. The electrodes were compressed under a range of pressures (1.4 to 60 kPa). It was found that below 40 kPa, the presence of bismuth increases the initial capacity. At all pressures, bismuth enhances the rate at which the capacity develops during cycling. It is suggested that reconnection of the separated agglomerates of lead dioxide is the key factor in restoring the capacity of the pulverized electrode. Electron micrographs reveal that there are two essential types of contact in the positive material: (i) 'micro-contact' between individual irregular-shaped or individual needle-like crystals, to form the agglomerates; (ii) 'macro-contact' between individual agglomerates, to form the skeleton of the positive mass. Bismuth encourages the growth of fine needle-like crystals on the surface of the agglomerates. These crystals spread out and inter-weld to form 'bridges' between the agglomerates thereby consolidating the porous mass of the electrode [131].

Bismuth addition (up to 0.06 wt%) to negative plates of lead-acid batteries improved the charging ability under partial state of charge (PSoC) cycling especially, for the low and narrow PSoC windows experienced in 42V power net automobile and HEV applications [132]. In order to provide the beneficial effect, the bismuth must be distributed uniformly in the negative plates.

Capacity and cycle life of VRLA cells containing a few ligninosulphonates (Vanispers A, Maracell XC-2, Lignotech-D-1380), Kraft lignins (Kraftplex, Indulin AT) and synthetic organic additives like Lomar *B, BNF and GKD,* in the negative plates have been evaluated.[133]. The negative active material (NAM) undergoes a progressive reduction in surface area resulting in a dense structure with low porosity. All the materials increased the capacity but considerable differences were observed in the capacity/dosage level responses. All the organic materials increase the surface area of the active material, and the surface area increases with concentration. The additives produce a morphology composed of small equiaxial grains with high porosity and surface area during cycling. The capacity and cycle life data show that each additive has a different optimum concentration in the plate. There are also indications that considerably higher concentrations of organic additives can be used in valve regulated batteries than is customarily used in flooded systems.

Chapter 7

CURING

Curing refers to the careful drying of the wet pasted plates during the manufacture of lead-acid batteries. Some manufacturers would like to call it "hydrosetting". Two curing processes are in vogue in the industry:

i. Low-temperature and
ii. High-temperature curing.

Both processes are conducted in high humidity environment.

Plates cured under low-temperature conditions, have mainly unreacted α-PbO and β- PbO, tribasic lead sulphate ($3PbO.PbSO_4\ H_2O$ = 3 BS), and a small amount of unreacted lead (free-lead). The level of β-PbO in the starting oxide has no major effect on either the phase composition or the morphology of the 3 BS cured material. Curing under high-temperature conditions results mainly in unreacted α-PbO and tetrabasic lead sulphate ($4PbO.PbSO_4\ H_2O$ = 4 BS), with a small amount of free lead.

The development of 4BS during high temperature curing has been explained in detail by Lam [134]. It can be defined in terms of four main features, namely, induction time, completion time, growth rate and maximum yield, the first three factors being influenced strongly by temperature, humidity and plate thickness, while the maximum yield is limited by the quantity of sulfuric acid solution that is used in the paste formulation. Research in the CSIRO laboratories [135-136] have identified and quantified a further determinant of 4BS production, namely, the level β- PbO in the starting oxide. It has been found that the induction and completion times are decreased, and the 4BS growth rate is increased, when the leady oxide contains up to a certain amount of β- PbO By contrast, the yield of 4

BS is unaffected. In another words, the development of 4 BS can be accelerated, but is yield is unaffected, by the presence of β- PbO in the precursor leady oxide.

According to Lam [134], there are no published data that show the effect, if any, of β-PbO on the plate formation, the initial capacity or the cycle life of batteries. The efficiency of plate formation is dependent upon the type of cured material (i.e., 3BS or 4 BS), the specific gravity of the forming acid, and the temperature and applied current profiles. The initial capacity and cycle life of batteries are also influenced by nature of the cured state, composition of the formed plates, acid concentration, temperature, and discharge rate. Battery cycle life is further affected by the type, composition and morphology of the grid alloys, the battery design (e.g., compression), as well as by the depth-of discharge. For batteries produced via the high temperature /high humidity curing route, it has been shown that 4 BS cured plates prepared from the optimum β-PbO content give longer service lives than plates in batteries produced via the conventional curing procedure.

The cured plates have sufficient strength for handling in the subsequent unit operations. This is achieved by the creation of amorphous materials, which could amount up to 20% of the dry paste weight. X-ray diffraction analysis of a cured plate would reveal about 10-20 wt. % of amorphous material. Leady oxide (gray oxide of lead produced in attrition mills or by Barton Pot Process) which is the starting material for the preparation of the plates, contains up to 30 wt.% metallic lead as fine particles (free lead in the battery parlance). During the paste mixing operation, the free lead begins to oxidize at a high rate producing the amorphous material. It is this amorphous material that acts as binders between the large tribasic or tetrabasic lead sulphate crystals, thus imparting the green strength to the dry pasted plates. Proper curing assumes importance in light of the fact that the sulphate crystals and oxide particles tightly glued together is the key factor for easy formation of the plates and a battery with longer service life [134].

Aidman [137] has reviewed the mechanism of expander action with special reference to the role of its organic constituent and the method by which this constituent is derived from lignin. SYNTAN BNF is proposed an alternative material for application as the organic constituent. BNF also increases the overvoltage for gas evolution, which could be of special interest for valve-regulated lead/acid batteries. BNF is compatible with alpha-oxy-naphtholic acid (AONA), an inhibitor of lead oxidation.

Chapter 8

ABSORPTIVE GLASS MAT SEPARATOR (AGM)

In 1967 work was begun on the research and development of small, cylindrical lead/acid cells containing spirally-wound electrodes using a separator material consisting of microfiber glass paper, now generally termed 'absorbent glass mat' (AGM). [138]. The tremendous success and wide commercial acceptance of valve-regulated lead/acid batteries for stand-by power applications over the last two decade have revolved round the manufacture and use of AGM. The glass fibre material's exceptionally small diameter (0.5-1μm), high porosity (95%), zero contact angle, high anisotropy and good compressibility makes it significantly different from conventional flooded separators. The electrolyte is retained and the oxygen transported within these nearly saturated recombination systems. Culpin [139] reviews critically some of the present methods for characterizing the glass-microfibre materials and suggests improved testing methods that reflect more accurately the subsequent effects on battery performance.

The fluid-transfer properties of the separator play a critical role in VR batteries. These properties are of particular importance in a VR lead-acid battery since the sulfuric acid is an active component of the battery reaction; the acid is depleted as the battery discharges. In a flooded lead-acid, the function of the separator to deliver acid is less significant than in a valve-regulated design. Zguris [140] discusses issues with regards to fluid-transfer properties of AGM.

Batteries manufactured using three different types of separator, but otherwise identical, were cycled to failure and the failed batteries were then subjected to a series of tests which included measurements of electrode potentials, compositional analysis of active materials, separator paper saturation and stratification measurements [141]. Results indicated that separator paper type significantly influences battery performance. It is believed that subtle changes in the separator structure caused changes in the rate at which imbalances between the cells in each test battery developed, which led to subsequent changes in failure.

Chapter 9

FORMATION

Plates cured satisfactorily are "formed" in dilute sulphuric acid, i.e., electrochemically converted to the respective active materials, namely, lead dioxide (chocolate brown colour) in positive plates and spongy lead in negative plates. This is carried out in a dilute solution of sulphuric acid usually 1.050 to 1.060 specific gravity (tank formation) made from pure battery grade acid. Plates must be fully submerged in the acid. Plates of opposite polarity alternate with each other having enough uniform space between them so as to avoid direct contact with each other. Plates of the same polarity are temporarily spot-welded to a lead busbar using a lead-arc brazing machine. Nowadays, tackless formation equipments are also available and can be used. Several tanks can be connected in series depending upon the voltage rating of the rectifier. The number of plates connected to in one tank depends on the current rating of the rectifier and also the dimensions of the tank.

After allowing a soak period of 1 to 2 hours, the positive pole of a rectifier (DC source) is connected to the positive plate busbar of the formation tank which may be of hard rubber or any plastic material or of wood with lead lining. DC current is passed through the tanks taking into account only the number of positive plates per tank. Polarity should be checked with a voltmeter within the first hour of formation to ensure proper connection. Formation is carried out at current density of 0.2 to 0.5 A dm^{-2} (both sides of the plates reckoned) over a period of 24-40 hours depending upon the weight of the dry unformed paste, thickness of the plates, . There are three main criteria by which one can decide the time at which the formation is to be terminated.

a) Freely gassing plates
b) "Clearing" of the plates (colour of the plates) particularly the positive plate with its chocolate brown colour.
c) The cadmium voltage readings.

Although the gassing starts even at midway of the formation process, plates gas profusely at the end. The pellet areas of positive plates will be free of any white patches (no surface lead sulphate) around the grid member. If small streaks of white material appear at the center of pellets, one need not try to remove them by continuing the formation since this may result in over formation and shedding. The presence of such slight amounts of lead sulphate (white streaks) rather helps in improving the life of batteries.

Chapter 10

CADMIUM VOLTAGE READINGS

These readings are important indications as to whether the formation is completed or not. The constancy in positive and negative cadmium potential readings indicate that the formation is completed. Since the cadmium readings taken on positive plates reach constancy several hours before the complete clearing it is better to take cadmium potential readings after the plates have "cleared".

The values for positive cadmium potentials range from + 2.30 to +2.45V and the negative cadmium voltage from –0.05 to –0.20V. Cadmium electrode is nothing but a cadmium rod or stick of 6-10 mm diameter and 50-100 mm length with a wire securely soldered at one end, the other end being covered with a perforated insulating material like rubber or PVC. Prior to use, the stick is seasoned in dilute sulphuric acid of proper specific gravity for 20 to 30 hours. To measure the potentials the cadmium electrode is attached to one of the leads of the voltmeter preferably a central zero voltmeter. If the meter reads in only one direction, then the cadmium electrode should be attached to the negative lead of the meter. It is then dipped into the forming electrolyte without touching the plates. Contact must then be made between the lead of the voltmeter will then be a measure of the potential between cadmium and positive plates or cadmium and the negative plates as the case may be. At the final stages of formation, the potential of the negative electrode shifts in the negative direction.

A word of warning is necessary with regard to the conclusions that should be drawn from the cadmium readings. Temperature effects have an influence on the cadmium negative readings at the end of formation. A high temperature reduces the voltage between the negatives and cadmium. Similarly, certain metallic impurities that may become deposited on the negative plates will reduce the cadmium reading to such an extent that it may be practically zero. When not in use for a long time, the electrode should be washed and kept dry.

There are three methods of formation to manufacture a VR battery [12, 142].

i. The plates can be tank-formed, assembled into cells and electrolyte can be filled under controlled condition after which the battery can be charged.
ii. Unformed plates can be assembled into elements in the containers (jar formation) in flooded condition, formed and processed further. The excess acid may be dumped.
iii. Unformed plates can be assembled into elements. The cells may be filled under controlled conditions with the correct volume of the electrolyte, formed, without the necessity of dumping the excess acid

In the case of first procedure, dry-charging of the plates after formation is advised. Of the three processes this happens to the most expensive, but it will provide the most reproducible product.

Batteries processed as per the second procedure are too wet initially to effectively recombine the gases, but they eventually dry out to give efficient recombinant batteries. This procedure also proves to be a cumbersome and messy one in that the acid specific gravity has to be adjusted and the excess electrolyte is to be dumped.

The last procedure is the most difficult to control and requires expert supervision. It is essentially a jar formation in a starved electrolyte condition. Plates for AGM batteries are normally formed to 70% while those meant for gelled batteries are formed to a lesser per cent.

The current required for formation is about 0.2 to 0.5 A dm^{-2} over a period of 24-40 hours, the more the formation time, the more efficient is the conversion efficiency.

The best procedure for a small manufacturer is to follow the first mentioned procedure, namely, tank formation combined with subsequent dry charging operation. The dry charged plates should be securely stored in air-tight containers until they are used.

With the type of controlled absorptivity AGM (CAAGM) separators (e.g., Hovosorb II separator, Trade name of Hollingsworth and Vose company) [143-146] significant cost savings can be achieved if such separators containing hydrophobic organic materials are used. The benefits of using CAAGM are as follows:

a) Unformed elements can be simply filled and dumped after formation ("fill and spill technique") without recourse to dry charging.
b) No necessity for dry-charging and the concomitant generation of wash water which requires costly treatment.
c) The excess acid during formation will prove to be a boon in better thermal management.
d) It avoids costly and time-consuming controlled filling of electrolyte.
e) It also helps in decreasing the potential for dendrites growth [147].
f) More uniform products are obtained with instantaneous recombination.

Jar formation of VRLA batteries poses problems and this process is less appropriate for large capacity batteries. General guidance for jar formation technique is given by Weighall [142]. The circumstances requiring jar formation are listed by Weighall: Cylindrical battery design; thin-plate, prismatic battery design; battery with a low height (L) plate-spacing (d) ratio, viz. L:d < 100; large separator fringe area; high separator grammage (>2g Ah^{-1}); high surface-area separator.

On the other hand, jar formation is not advised for the following circumstances: tall batteries, large, high-capacity batteries, deep cycling long-life batteries, batteries with high L:d ratios (>100).

"One battery using one circuit" concept is strongly recommended by Bitrode for formation cycle [148]. This concept allows maximum control, flexibility, and precision in the formation process.

In the manufacture of valve regulated lead-acid battery bottlenecks occurring on the production line can result in varying delays. Lawrence [149] describes the effects of different soaking times between the 'acid fill' and 'formation' stages of production. They have used X-ray diffraction (XRD), BET surface area analysis, scanning electron microscopy (SEM) and electrical testing to examine the compositional and morphological aspects of the positive active material just prior and after formation. Paste compositions are effected by the time period between acid fill and formation. However, electrical test results and SEM examination of formed pastes suggest that soaking time is unlikely to affect the resulting battery performance.

Chapter 11

ASSEMBLY OF VR BATTERIES

In the battery assembly, negative plates flanking the positives is normally employed. Assembly of gelled and AGM VR batteries differ in that the former can be assembled in the usual manner, i.e., the groups can be separately burned and interleaved after which separators can be inserted. However, in the case of AGM batteries, considering the very fragile nature of the separator material, it is impossible to assemble groups and then insert the AGM separators. Hence the plates have to be stacked one over the other with one or two layers of AGM separator in between them. The burning of connector strap / busbar and pillar posts can be completed simultaneously for both the groups manually or by using Cast on –Strap machine. (COS).

For group burning, the welding sticks/rods should preferably be made of a lead-tin alloy for the Pb-Ca plates and the same alloy as the grid material can be used for the low antimony batteries. Here, the high temperature of the welding torch is likely to oxide the calcium and thus produce the dross creating an ugly finish. Some manufacturers prefer to use a hood or housing into which a supply of some inert gas (e.g. nitrogen or carbon dioxide) is admitted. This may help to reduce the drossing out of calcium and thus improve the finish of lead burning.

The assembly of tubular type VR lead-acid battery differs from that of flat type in that almost all manufacturers use the GE technology. However oxide-holders of square design can be used with AGM separators. With the work of ALABC, flat spine concept has received much attention and is well suited for AGM assembly. Either the loop-over-the-partition (LOP) (also called the "up-and-over" connectors) or "through-the-partition" (TP) connection for inter cell connectors can be employed.

Chapter 12

CONTAINERS AND COVERS

The containers need not have the mud space (bridges in the bottom) partly to make up for the reduced volume of electrolyte and partly because no sludging of active masses is occurring. The covers may be a single piece top cover with provision for fixing the pressure vents or there may be two covers. The first one with vent holes is sealed to the container top. The second one is fitted to the first cover for giving a smooth outer finish. The second cover should naturally have one or two small holes to allow the excess gases to escape.

Chapter 13

PRESSURE VENTS (SAFETY VALVES)

The pressure vent design may be of the push-on-cap type or may be the threaded type depending upon the design of the cover. The principle is the same. A special rubber cap covers a cylindrical exhaust tube. As the internal pressure in the battery reaches the limit, the valve lifts (opens) to release the accumulated gases and when it attains a predetermined pressure, the valve closes and remains so until the internal pressure again exceeds the vent pressure.

Typical pressure range employed for this purpose is 20-28 kPa (3-4 psi). However, valves with vent pressures from 3.5-350kPa (0.5 to 50 psi) are also in use [2]. The Gates Energy Products cylindrical cells are designed to vent at very high pressure of 275-415 kPa (40-60 psi) [150], the Yuasa NP series at 20-50kPa (3-7 psi) [151] and the Dryfit Sonneschein batteries at 10 kPa (approximately 1.5 psi = 0.1 bar). A few vent designs are shown in Figure 3. The reader can see references [33, 152-155] for vent designs.

In some instances, blocks of cells had become stuck together by the combined action of heat and expansion. By employing a ribbed box and a steel shim spacer assembly, it was possible to prevent the center cells from becoming as much as 11° C hotter than the end cells and hence the concomitant premature battery life was eliminated. To reduce the transmission of oxygen and moisture through plastics, a poly (vinyl chloride) jar and cover were introduced. An extensive programme of accelerated life testing was undertaken. [156]. This resulted in the development of the 6mm positive grid with its unique six-segment design for horizontal growth control. Instead of the horizontal growth transmitted continuously on the two sides of the plates, the plate grows in six segments to a lesser degree. A number of severe problems were traced to poor jar-formation procedure. The problems included heavily sulfated plates, remarkable variation in specific gravity and free acid, enormous variation in capacity, and poor and widely differing levels of compression. Therefore, to avoid these problems, C and D "Liberty 2000" product uses the following three techniques:

i. tank formation
ii. Controlled compression
iii. Unique pressure plate system

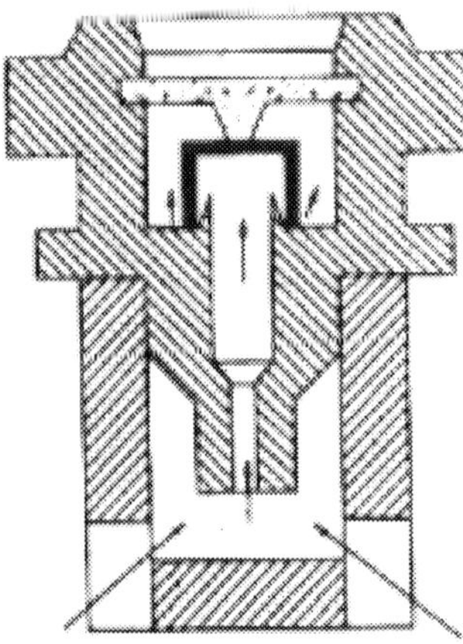

A

K. Eberts, Power Sources 2, (ed) D. H. Collins, Pergamon Press, London 1970, p. 77

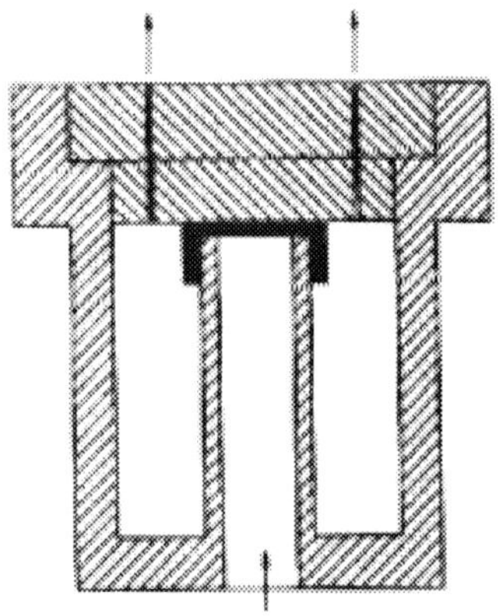

B

T. Yamada et al., J. Power Sources, 38 P1992, 124.

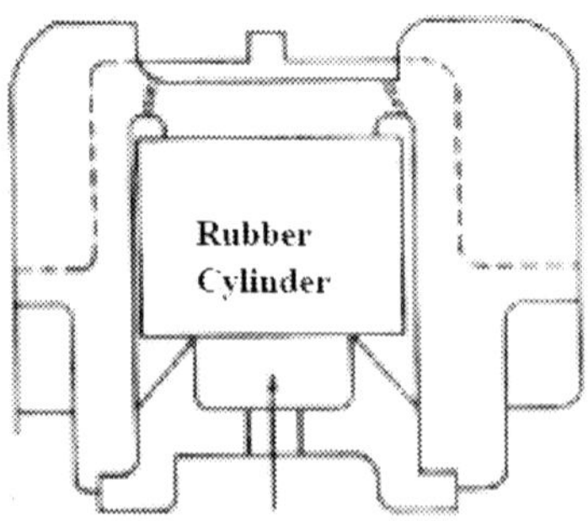

C

J. Szymbonski and B. Burrows in Power Sources 9, (ed) J. Thompson, Academic Press, London 1983, p. 116

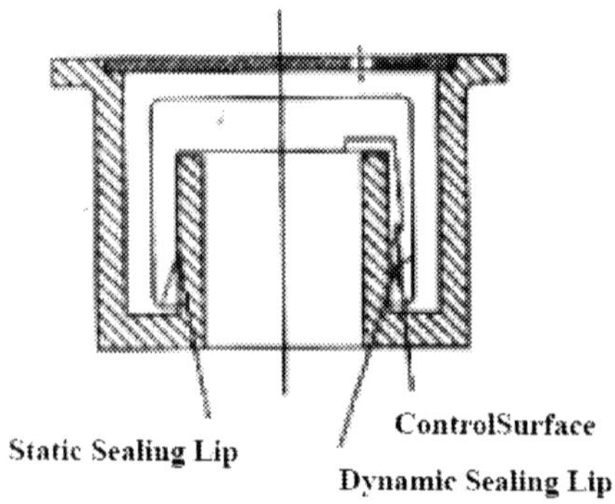

D

Brochure on "Ventseal," Bursak + Shamban GmbH & Co., Handwerkstrasse, 5-7, D-70565, Stuttgart, Germany, 1995.

Figure 3. Valves.

Chapter 14

ASSEMBLY OF GELLED VR BATTERIES

The assembly of such a battery is simpler and follows the normal procedure. After finishing the assembly, the elements should be fixed to the cell walls by applying some adhesive (epoxy cement) so that the whole element structure is not disturbed during transportation or under circumstances where the battery is likely to experience vibrations. After filling the battery with gelled electrolyte and sealing with covers, a few labour and energy consuming conditioning cycles are to be given.

Cycle life tests were carried out to evaluate the influence of safety valve pressure on valve regulated lead/acid batteries under deep cycling applications[157]. Batteries were cycled at 5 hour rates at 100% DOD, and the safety valve pressure was set to 1.08 and 2.00 bar, respectively. The batteries lost 248.3 g and 235. 3 g of water for each case after about 1,200 cycles, but the cyclic performances of the batteries were comparable. Most of the gas of the battery during discharging was hydrogen, and the oxygen concentration increased to 18% after 3 hours of charging. The micro structure of the positive active materials was completely changed and the corrosion layer of the positive grid was less than 50 mm, regardless of the pressure of the safety valve after cycle life tests. The cause of discharge capacity decrease was found to be water loss and the shedding of the positive active materials. The pressure of the safety valve does not give little effect to the cyclic performances and the failure modes of the gelled electrolyte valve-regulated lead acid batteries.

12 V/25 Ah gelled-electrolyte valve-regulated lead-acid batteries were evaluated for their performance in relation to the AGM valve-regulated and flooded counterparts at various discharge rates and temperatures between - 40°C and 40°C [158].

Although the performance of the gelled-electrolyte valve-regulated battery is similar to both the AGM valve-regulated and flooded lead-acid batteries at temperatures above 0°C, it is superior to both the flooded-electrolyte and AGM valve-regulated lead-acid batteries at temperatures between 0°C and -40°C, which is the characteristic that makes it attractive to expand the application regime of VR lead-acid batteries. The corrosion rate of the positive grids in the gelled-electrolyte battery is also lower when compared with those used in the flooded-electrolyte and AGM configurations.

Chapter 15

ASSEMBLY OF AGM VR BATTERIES

In the assembly of AGM Batteries care should be taken to closely control the following parameters.

i. plate thickness, porosity and active material weight.
ii. Separator thickness before and after compression.
iii. Interplate distance (plate pitch) and
iv. The ratio of active material to electrolyte volume which automatically determines the saturation level of the separator used. For a better performance the mass balance of PAM:NAM varies between 1.3 to 1.75 and the volume of electrolyte varies depending on the application [77].

After cleaning the lugs for good contact during the lead burning operation, the required plates are stacked either manually alternating with each other with two layers of AGM separator in between them or plates can be wrapped by commercially available automatic wrapping equipment designed for this product. The separator should extend on all the sides except the bottom by at least 2 to 5 mm so that shorting can be avoided. It helps in increasing the surface area of the separator and so the volume of acid. This is possible only if the plates are about 5 mm lesser in width than the internal dimension of the container.

Wrapping technology up to recent times had been manual.

The methods of wrapping practiced can be classified [159] as follows.

A. Zig-zag wrapping: Separator of definite length is folded beforehand in two directions such that it can be folded in a zig-zag manner, here both positive and negative plates are manually placed.
B. Single piece separators are wrapped around each plate (both positive and negative) and then stacked to elements.
C. Automatic wrapping and stacking machine. Two machine types are available (i) wrapping of positive only and stacking them together with negative and (ii) wrapping of positive and negative and stacking. Plates of width 40-215, height, 60-260 and thickness 1 to 6 mm can be wrapped using such machines. Separator thickness may vary from 1-35 mm.

The arrangement of plates and separators must be pre compressed before being made into an element with connector straps and posts. [160].

Without disturbing the arrangement and without damaging the highly fragile AGM separators the element is to be inserted into the group burning racks and connecting straps/busbars and suitable pillar posts can now be burnt on to the grid lugs. A cast-on-strap (COS) machine can be used with advantage for such assembly.

The separators must be firmly in 100% surface contact with the plates [71, 161-164]. This is achieved by compressing the separators between 10 – 30% of their original thickness normally expressed under a compression of 10kPa (1.5 psi) [164]. For example if the declared thickness is 2 mm the separator must be compressed so that the thickness decreases by 0.2 to 0.6 mm. While designing a jig/comb for assembly of such batteries this reduction in thickness should necessarily be taken into account. Excessive compression beyond 30% should in any case be avoided since this will reduce the void volume and thus adversely affect the recombination efficiency (RE) [71,161-162].

To avoid such critical issues, industry is developing a stiff separator which exhibits greater thickness when wet. The increased stiffness makes the separator suitable for standard line machine assembly. Although pre-compressed separators have been sold for years, recent times have seen a tremendous increase in R and D of such materials since there is a good potential for increasing the manufacturing output [146,165]. Because of such separator availability in standard sizes cut to over-size relative to the plate dimensions and assembled flat without wraparound are possible [144].

Holden [166] describes an alternating C-wrap approach (as opposed to the conventional U-wrap approach) to avoid misplaced and damaged AGM separators which are common. This removes any possibility of shorts.

Chapter 16

FILLING OF VR BATTERIES WITH ELECTROLYTE

There is a dearth of electrolyte in VR batteries. To compensate for this a little higher final specific gravity acid is used (1.280 to 1.360). If the plates are only partially formed, a lower specific gravity acid should be used for filling (1.230 to 1.250). The volume of electrolyte is about 9 to 10 ml per Ah at 20 hour rate, but some authors state a range of 10 to 11 ml per Ah [38]. The batteries with lesser volume of electrolyte would show recombination even in the initial few cycles and the latter type may require a dry-out time for the recombination to start. [12].

The electrolyte should contain about 10g of ortho-phosphoric acid per litre of the electrolyte if the grid alloys for the positive have no antimony. This will fight against the electrochemically inconvertible lead sulphate that will be formed in the absence of phosphoric acid in lead calcium batteries [29,167-169). This will also improve the rechargeability after a deep discharge. In charged batteries, the phosphoric acid is contained in the active materials of the electrodes. During discharge sulphuric acid is replaced partially by the phosphoric acid. This phosphoric acid helps in the conductivity during a deep discharge and so the battery remains rechargeable without any difficulty. To avoid hydration shorts, most manufacturers add sodium sulphate to the electrolyte (up to 10 g litre^{-1}).

Chapter 17

FILLING OF AGM BATTERIES WITH ELECTROLYTE

Although with the advent of a controlled absorptive AGM separator, precision filling is not required, for the conventional AGM separator, precautions for maintaining a definite ratio between the electrolyte volume and active material should never be overlooked. The calculations are based on the determination of porosity of the plates and arriving at the total volume available for the electrolyte. A saturation level of 90 to 95% of the pores of the separator is the best for efficient recombination [69,71,161,163]. As the saturation level slowly increases as the addition of acid proceeds, the electrolyte connects small pores and leaves the larger pores open and free to the passage of gas. Initially the matrix will be saturated with acid and therefore the gas phase mass transfer of oxygen to NAM will be impeded with a resultant increase of internal pressure and venting. Eventually a balance will be reached when there is sufficient free volume in the porous matrix to permit the rate of mass transfer and subsequent recombination of oxygen at the NAM to keep pace with the rate of oxygen generation at the PAM. The latter is determined by the over-charge current which is in the order of C/1000 [163].

The water loss during the entire life of good VR batteries is negligible. The value depends on the float voltage, temperature of operation as well as the alloy used in the design of VR batteries. For example Peter [163] quotes a value of 40ml for 10 years for telecommunication and UPS batteries floated at 2.27 V/cell (25°C) and gives a figure of 99% recombination efficiency.

Although the recombination is dependent mainly on the volume of electrolyte it also depends on the partial pressure of oxygen as well as the type of separator; it is little affected by the electrolyte concentration. The electrolyte concentration has, on the other hand, a significant effect on the electrical output of the VR battery [71]. The cells are then sealed with cover as usual.

Numerous studies on gelled electrolyte batteries are available in the literature [170-172].

The special requirements of separators used in the manufacture of gelled valve-regulated lead/acid (VRLA) batteries start at the cell-processing steps, where the separator must assist, through its characteristics, the formation of an optimal gel structure. In a gelled VRLA cell, conditions exist that pose additional demands on the separators. In this type of cell, the separator is not a by-stander, but an integral link in the chain of events that comprise the establishment of an efficient oxygen-recombination cycle. Ferreira [170] reported the need for separator materials of very high porosity., as well as the need for optimal pore-size distributions. An examination is also made of the effect of the surface characteristics of the separator on the formation of gel micro-channels and, in term, the efficiency of the oxygen cycle.

Chapter 18

FILLING OF GELLED VR BATTERIES WITH ELECTROLYTE

The electrolyte is prepared by adding 4-7% of a very fine fumed silica of large BET surface area to 1.280 – 1.300 sp.gr. sulphuric acid while continuously stirring the solution with a high speed mixer [173]. The particle size of the silica is in the range of 5 to 20 microns. The gel thus prepared should have a good thixotropic property so that it can flow through the spaces between the separator and plates and fill all the voids before it can set to a gel.

Granular silica [37] is advocated to increase the life as well as capacity per unit weight of PAM.

The filling of the gel electrolyte is done under vacuum [174]. An arrangement with cells inside the vacuum chamber with transparent windows for observation and with nozzles positioned exactly above the filling holes of the cells is a convenient device. The nozzles should have opening and closing arrangements with stop cocks and they should be connected to the reservoir of the gelled electrolyte which is kept agitated during the addition.

Calculated amount of the gelled electrolyte is added to each cell under vacuum (10 to 30 mm of mercury) for at least 10 minutes so that all the internal space is taken up by the gelled electrolyte. After the introduction of the electrolyte, the vacuum can be released. The incoming air improves the contact between the gelled matrix and the plates when the atmospheric pressure is brought to bear upon the surface of the electrolyte. The filled battery is now allowed to settle for a few hours or until extra liquid forms on the top during ageing. This extra liquid should be discarded by tilting the battery. Then the battery is sealed as usual except for the pressure vent.

The electrolyte should have 20-35 gramme per litre of concentrated o-phosphoric acid per litre of electrolyte to compensate for the absence of antimony. This is supposed to give a better positive plate performance [175].

Chapter 19

CONDITIONING CYCLES

After filling with the electrolyte, the cells are given a charge/freshening charge so that they attain a terminal voltage of 2.45 2.55V per cell (VPC) and are maintained at this voltage level for about 2 hours. They are then sealed with proper pressure vents.

This part of the manufacturing step is necessary to meet the specification of the declared initial capacity. Here the concept of recycling the energy can be advantageously employed for charging a series of batteries using another set of batteries (naturally of higher voltage rating) which require being discharged in this conditioning schedule.

Automatic life cycling equipment will lend itself for performing the conditioning cycles. At least 2-3 conditioning discharge cycles should be given.

Finally the individual batteries should be tested for high rate discharge to check for any defect in the assembly. For example, the batteries may be discharged at 3 C rate, i.e., for a 100Ah battery (the current for discharge is 300A) for 7-8 seconds at the end of which the terminal voltage should not be lower than 1.66V per cell.

Conditioning of AGM Batteries: The batteries can be discharged after fixing the pressure vents in place. The rate of discharge can be anywhere between C/22 to C/5. The cells are now recharged after they reach a terminal voltage of 1.75 volts per cell (VPC) at 2.30 VPC constant potential (CP) charge or at C/20 constant current (CC) charging scheme until about 10 to 15% extra input is attained. In the case of CP charge it will take at least 12 to 16 hours and in the case of CC mode 22-24 hours. To reduce the time of cycling a 2-step charging and a C/5 rate of discharge can be employed.

During the charging process, before sealing with pressure vent the voltage has to be taken to a value of 2.55 VPC. Conditioning of GE VR batteries: After sealing the cover to container, the batteries are subjected to several (about 5) cycles to bring them to the actual recombining state. The only precaution to be observed is that the terminal voltage should never be allowed to go beyond 2.45 VPC in any circumstances. This is to control the gassing to a level that produces the desired micro cracks in the gelled matrix. [176]. Excessive gassing will cause large voids / cracks, which separate the gelled matrix into unconnected pieces thus reducing the capacity and the ability for gas recombination. Compared with AGM batteries this part of the process is time consuming and increases the cost. Nevertheless, it is a must. After giving about 5 cycles the batteries are sealed and they are ready for shipment.

Chapter 20

ADVANCED LEAD-ACID BATTERY CONSORTIUM

The Advanced Lead Acid Battery Consortium (ALABC) [177-178] was set up in March 1992 to manage a four-year Research Plan with a fund of US $ 19.3 million for the development of a high performance EV lead-acid battery that will serve a significant share of the EV market in the short to mid-term. The ALABC is managed by the International Lead Zinc Research Organisation (ILZRO) and is a partnership organisation among fourteen largest lead producers, twelve battery manufacturers, electric utilities, motor manufacturers, charger and coupling manufacturers, power-train suppliers, controller/electronics manufacturers and EV trade organisations. The membership currently stands at 60, hailing from 13 countries [87].

By using high tin (1-1.5%) lead alloys [83-84,87] for light-weight grids and spines of special shapes, the coefficient of use of positive active material (PAM) has been raised to 50% at 3 h rate which may result in battery with specific energy of 42 Wh kg^{-1}. Combined with such a design, special addition like porous polypropylene can enhance the special energy to about 47-50 Wh kg^{-1} [87].

It has been established [179] that when the density of the active mass falls below a certain critical value (-3.4 g cm^{-1}), the structure of the PAM is very difficult to restore upon plate charge and some regions of the plate are excluded from the current-generation process . As a consequence, the plate capacity declines. To avoid this situation, the pulsation of the PAM should be restricted. This can be achieved by confining the plate within a given volume (for example by placement of the PAM is tubes) or by subjecting the battery plate in the action of a force that would counteract to pulsation. Takahashi et al. [180-182] have determined the changes in capacity, electrical resistance and thickness of the position plate on cycling when a force of 0.2 or 1.0 kg cm^{-1} is applied to the cell (Figure 20), The cells under higher pressure exhibit smaller increase in thickness, lower ohmic resistance, higher capacity and longer life. When the pressure over the plates is increased from 0.2 to 0.4kg cm^{-2}, the cell life is increased by a factor of two. The ability of the plate to pulsate on cycling is restricted to about 15%. Hence, the life of the battery is very sensitive to changes in the PAM density.

Chapter 21

ELEMENT COMPRESSION

The element compression (Expansion of PAM) to about 40 kPa has been shown to improve the capacity and the life from 200 to more than 700 cycles [180]. Under such compression, the AGM is likely to collapse and hence a variety of different separator/compression systems is currently under development. Compressive forces applied in the direction perpendicular to the plane of the plate prevents swelling of active material and counteracts the tendency of PAM to increase in electrical resistance.

The spring characteristics of a typical recombinant battery separator mat (RBSM) material used in valve-regulated lead-acid (VRLA) batteries have been monitored at several stages during repetitive deep-discharge cycling service (C-3/3, 100% DoD)[182]. Through the controlled application of a range of compressive loads, accurate plots of separator thickness vs. compressive force have been recorded for both dry and acid-saturated material. From subsequent analysis, several important properties had a second stage of crush that can be permanent. Saturation with sulphuric acid solution produces a discernible shrinkage of the material. Thus, for a given than a dry equivalent. This has important consequences for the manufacture of RBSM VRLA cells is not an effective strategy for attaining the levels of compression required to prevent significant expansion of the positive plates. The difficulties of setting, and then maintaining, adequate levels of plate group compression are further underlined by the promising performance of a constant-compression RBSM VRLA cell. By means of a feedback-controlled system that applies a pressure of close to 40 kPa, this first-of-its-kind cell has displayed relatively stable cycling behaviour.

Chapter 22

FAST CHARGE

Very fast charge has been met with great success and results in higher life [183-186]. A survey by the Cominco Research Team [184] has shown that thirty commercially available varieties of lead-acid batteries are capable of being recharged to 50% in 5 minutes, 80% in 15 minutes and 100% in 30 minutes. In this respect the performance of VRLAB is better than flooded SLI batteries. Positive active material from conventionally charged plates are characterized by larger particles and numerous big pores. In the rapidly charged plates observation showed that no large particles' pores are voids, whereas the conventionally charged plates exhibited 2 m^2/g surface area of PAM; those charged at high current exhibited surface area value of 3 m^2/g even after 900 cycles. The PAM in the latter case expanded only slowly and consequently less pressure was exerted on the separator and the negative plate, thus reducing the risk of "soak through" shorts in the separators and NAM densification. Another work has shown the dramatic effect that fast charge has on life cycle. When 12V/50Ah spirally wound VRLAB were tested under 10h and 15 minutes charge regimes, the conventionally charged batteries could give only 250 cycles (to 80% of initial capacity) whereas those under fast charge regime could give about four times more cycles [184].

The benefit of such a fast charge is that an EV powered by VRLA can give 1640 km in 24 h at more than 70 km/h speed with intermittent fast charge, the normal range being 160 km/charge [84].

The VRLAB has already met all but one (specific energy) of the mid-term criteria of the USABC [84]. Batteries are also available commercially. 12V valve-regulated lead-acid batteries (spirally wound type) were subjected to conventional-charge and rapid-charge cycling tests [186]. The cycle life was 250 cycles for the conventional-charge regime and 1000 cycles for the rapid-charge regime. In conventional-charge cycling, the positive active material quickly expanded and developed a coralloid structure in association with lowered utilisation and integrity. In rapid-charge cycling, no coralloid structure developed and the expansion was smaller and much slower. Correspondingly, the particle size of the negative active material grew in both cycling tests, but at a much slower rate in rapid-charge cycling. With the expansion of the positive active material, the negative active material was compressed. In the failed batteries, about one-third of the negative active material in the centre of the electrode was compressed almost into a solid non-porous mass. This densification process also occurred at a much slower rate in rapid-charge cycling. At the point of failure, the discharge capacity of all test batteries was limited by the negative electrode, although it was limited by the positive electrode at the beginning of the cycling tests. The cause of failure for most of the batteries, regardless of the charging regime, was the occurrence of "soak-through" shorts caused by numerous minute lead dendrites formed in the separator. This might have been encouraged by the formation of shorter distances between the two electrodes, created by the compression of the separator as a result of the expansion of the positive active material.

A new AGM valve-regulated lead-acid battery with weight and volume substantially reduced, as well as a substantial improvement in its reliability, has been reported by Soria [187]. These batteries are more efficient, safe and supply energy economically. Battery specific power is practically related to electrode area, so that its increase, and therefore a thinner plate was used. Furthermore, it was necessary to achieve a similar reduction in the conventional glass microfibre separator. But such thin material should have impaired mechanical properties and can make the battery more prone to develop short circuits across the separator. Hence a new microporous polyethylene membrane was developed and tested, with excellent mechanical properties, high porosity and low pore size. This improved design is combination of non-woven glass microfibre and the polyethylene membrane. The partners in the development were: Batteries :Tudor (Exide Technologies); UPS Systems : MGE; Daramic : polyethylene membrane; AGM: Bernard Dumas; Fundamental research studies on very thin electrodes: Inorganic Chemistry Department of Cordoba University.

Chapter 23

NEW GENERATION AUTOMOTIVE BATTERIES

The battery employed in automobiles faces the possibility of a major change, in that the electrical systems planned for vehicles are expected to be far more extensive than hitherto, i.e., from a 12V system to a 36V system. These advanced electrical systems will require the electrical energy-storage element to operate at a partial-state-of-charge (PSoC) for most of its life and at unprecedented high rates (so-called 'HRPSoC duty'). The remarkable advances achieved with valve-regulated lead-acid (VRLA) battery technology for electric vehicles during the past 8-10 years will be of only passing value in overcoming the challenges posed by HRPSoC duty. The battery failure modes seen in HRPSoC operation are quite different from those faced in EV (deep-cycle) use. In summary, it has been established that in HRPSoC service, which batteries will have to undergo in future automobile electrical systems, VRLA batteries are plagued by a sulphation problem of the negative plate. Appropriate design of the plate grid can enhance the ability of the battery to deliver and receive charge at the required rate. Furthermore, the introduction of additives to the active material can delay the onset of gassing, and hence assist recharge of the negative plate [188-189].

Chapter 24

OPERATIONAL EXPERIENCE

In order to investigate the behaviour of valve-regulated lead/acid batteries in solar power applications, gel and AGM batteries were installed in different solar power systems [190]. Each system is divided into several groups and each group has the same battery type, the same loading and the same solar generator. The only difference is the charge/discharge strategy. A key result after 2 years of testing is that the charge strategies which are typically used today in the field cannot charge the batteries completely. However, if the batteries were charged intensively afterwards they returned to full capacity. This means that there is a problem of undercharging in the field. Improving the charge/ discharge strategy can, therefore, extend the service life and the energy turnover of VRLA batteries in solar power applications. Moreover, some existing VRLA battery types were modified with regard to the amount of electrolyte and phosphoric acid. These versions were investigated in the laboratory and are included in the field tests.

A low-cost, high energy-density lead-acid battery employing lightweight plastic grids has been developed and tested [191-192]. These grids were coated with electroless copper for 5-10 microns. Then the grids are electroplated with lead for a thickness of 50-200 microns. The grids electroplated with lead are subjected to Rapid Thermal Process (RTACRP) for coating tin oxide. The thickness of the coating varies from 5 to 25 microns. Such grids can offer highly improved performance with respect to specific energy.

A new concept of gelled-electrolyte-AGM-hybrid VRLA cells is being promoted by Shukla and his co-workers and such batteries are under field trials for solar photovoltaic applications and other stationary applications [193-194].

Chapter 25

FAILURES OF VRLAB — CAUSES AND REMEDIAL MEASURES

The electrochemistry of lead-acid battery by which energy storage and delivery occurs seems to be simple apparently. But the performance of the battery depends upon a number of factors such as cell design, raw materials used, various parameters involved in plate preparation, chemical composition/structure of the active materials and the conditions of battery operation. As a result, degradation of performance and eventual failure of the battery is a cumulative effect of many factors related to the above. Failure may be a catastrophic one or a progressive one. Valve regulated batteries faced many problems at the early stages of introduction. But these have been improved steadily and VRLAB has evolved into a mature technology at present. Potential causes of failure and factors that effect durability are discussed here. In addition to a precise manufacturing process, battery monitoring and surveillance is also important for reliability and durability of VRLAB.

Failure of lead-acid batteries can be categorized as given below.

1. Catastrophic failure and
2. Progressive failure

Catastrophic failure results from any one or more of the following reasons

1. Incorrect cell design and component selection
2. Poor quality control during manufacture
3. Abuse of the battery such as under-charging, overcharging
4. External and/ or internal damage.

Progressive failure originates from adverse changes in plate characteristics caused by the influence of manufacturing variables and service conditions such as

i. Nature of raw materials used such as grid material, leady oxide, sulphuric acid, separator, etc.
ii. Paste composition and pasting
iii. Method of curing.
iv. Chemical composition and structure of formed plate
v. Nature of charge and discharge
vi. Temperature of operation
vii. Depth of discharge, etc.

Main failure modes of valve regulated lead-acid batteries are

1. Grid corrosion and grid growth
2. Water loss
3. Thermal runaway effect/ dry out
4. Premature capacity loss (PCL)
5. Positive active material softening and sulphation
6. Negative plate failure

Positive Grid Corrosion

Although the normal predominant failure mode of any type of lead-acid battery is corrosion and disintegration of positive grids, it is particularly so in batteries using Pb-Ca alloy grids that have not been properly cast employing proper composition. The parameters that affect the corrosion rate are:

- Alloy composition
- Design of grids
- Casting conditions
- Impurities in the alloy
- Operating temperature of the cells and
- Charging voltage of the cells.

Because only a narrow range of Ca content in the alloy is acceptable in view of the facts given under alloys and casting, strict control on Ca content and casting conditions are required. The starting material for alloy preparation is essentially 99.97-99.99% pure lead (primary or secondary) with certain impurities within the limits.

The charging voltage of VR cells should be compatible with the electrolyte specific gravity to obtain optimum life [195]. The corrosion rate does not only depend on the charging voltage, but also on the recombination efficiency. The latter influences the positive plate potential which is the determining factor for corrosion along with the electrolyte specific gravity. In VR cells the positive plate experiences conditions which are more corrosive than in flooded cells. This is due to the depolarizing effect of the recombination reaction on the negative plate. For the same impressed terminal voltage, the current is increased and the voltage supplied to the positive plate is increased [196].

SOFTENING OF PAM

Valve-regulated lead-acid batteries undergoing intensive utilization of the positive active-mass suffer softening of this material, thereby reducing battery cycle-life. Investigations were carried out in order to understand the mechanism of softening [197]. The evolution kinetics of the β-PbO_2 were studied with very dynamic discharge profiles: different steps of the β-PbO_2 life were explored. The experimental results were compared and correlated with positive active-material used in electric vehicles (field tests). Scanning electron microscopy revealed a continuous growth of the β-PbO_2 microstructure throughout the cycle-life of the active mass (from 1 to 2 μm) which was confirmed BET surface area analysis of the active materials (from 4 to 2.5 m^2 g^{-1}). Powder X-ray diffraction revealed similar changes in the nanostructure (from 20 to 90 nm): the crystallites, as microstructure particles, grew with the number of cycles. These results also revealed the difference between the active material bound to the electrode and the β-PbO_2 crystallites which had lost contact with the plate; the latter being larger than the former (100 nm vs. 80 nm). The investigations confirmed that the capacity loss of VRLA batteries was strongly linked to a micro- and nano-textural evolution of the β-PbO_2. This finding supported the theory that there is a progressive expansion of the positive active-material, which leads to a decrease in the contact zones resulting in the shedding of the material. Strategies to overcome this effect were tested and it was found that electrolyte additives and electrical treatments could be useful in extending battery cycle-life.

LEAKS

Because VRLABs have to operate under pressure to facilitate "oxygen cycle", there are a few types of leaks of acid [79]. Thus if the sealing of cover to container is not properly checked during the manufacturing stage, or if the posts are not properly sealed to the cell covers or if cells are overfilled with acid, the electrolyte may leak out because of all these causes. However, these problems can be easily eliminated by improving the practices and by having strict quality control measures. For example, special coating can be applied to lead bushings to improve adhesion to plastic covers, each individual cell can be checked for the integrity of cover to container seal pneumatically and acid volume can be controlled to rectify the particular type of leaks [198]:

1. Many jar-to-cover leaks were found can be mostly eliminated by designing a tongue-and-groove adhesive seal, instead of heat seal. This design gives a seal path that is three times longer [156].
2. A composite bushing of rubber and lead-tin alloy which is glued into the cover and a weld that is made by a computer controlled, automated, proprietary machine is sure to eliminate post-to-cover leaks. A highly reliable, low pressure, umbrella valve (0.8-1.0 psi) can withstand higher pressure than Bunsen valves which cause problems of bulge, jar-to-cover leaks and sticking.
3. Many of the post seal-leaks were exacerbated by stress. Hence, an element suspension system was developed. A sturdy, injection – moulded, polypropylene component fits on the bottom of the element via lugs on the grids. There is a pocket moulded into the part that coincides with the positive, and this allows for growth. When cell is in service in the horizontal position, the element is supported by this component. A moss shield (that also serves as a top support) compliments the bottom support. Again it fits over the lugs prior to burning and in the horizontal position, provides a support at the top of the element. The net result is that, in the horizontal position, the whole weight of the element is suspended on the top and bottom supports. Thus all the stress is removed from the post seals and the jar-to-cover seal, yet there is allowance for growth into the pockets.

Charging of VRLA Batteries

Proper charging methods help a lot in obtaining good life from batteries.

Only constant-voltage or constant-potential (CV or CP) charging is recommended for VR batteries. Because of the high charge acceptance of the battery, the initial current can be anywhere between 0.2-0.3C amperes (i.e. 20 to 30 amperes for a 100 Ah battery, irrespective of the voltage) depending upon the impressed voltage and the state of charge (SOC) of the battery. The lower the charging current, the more will be the duration for obtaining full charge and vice versa. For example, after a full 2-3 hour rate of discharge the values will be as given in Table 3. The batteries can be charged at 0.20 CA only for about 2 h or until the terminal voltage reaches a value of 2.30-2.35V per cell; thereafter the charging mode should be changed to CP.

Model predictions of half-cell voltages, current, and gassing behavior are compared to experimental results of valve-regulated lead-acid (VRLA) cells to elucidate the charge mechanisms. Good comparisons of experimental Pb half-cell voltages with model predictions confirm the importance of liquid-phase Pb^{2+} transport early during charge. In the latter stages of Pb-electrode charge, limitations in the $PbSO_4$ dissolution rate are more important than those of Pb^{2+} transport and control charge behavior. Model predictions with an analogous mechanism for the PbO_2 electrode (i.e., involving dissolution and Pb^{2+} transport) were not consistent with the charge polarization behavior, since comparisons of experimental half-cell voltages with model results were poor. Satisfactory comparisons with a model that incorporates a solid-state PbO_2 charge mechanism confirmed the insignificance of dissolution and transport in PbO_2 charging. The good agreement between model and experimental current behavior during the constant-voltage portion of charge support our conclusions regarding the electrode charging mechanisms. Conditions under which to expect charging difficulties as well as improved charge regimes are also discussed.

Table 3. [199]

Current value and period for a CP Charge		Approx duration to attain 90-95% SOC
Current (amperes)	**Hours**	**Hours**
0.3C	2.5	15
0.2C	4	18
0.1C	8	20
0.05C	18	30

The model also predicted the characteristic double-peak gas-flow behavior seen when charging VRLA cells with constant current to a specific voltage lid. The first peak is due to the onset of significant amounts of O_2 generation even though the internal residual gas is dominantly H_2. The second peak is due to the onset of H_2 evolution [200].

Voltage Variation in Float Operation

It is well known that the float voltage of individual cells in a series string of VRLABs frequently differ significantly from each other, particularly during the early stages of service life [11, 79]. Although the applied voltage per cell is 2.3 VPC, there is large voltage distribution in the case of VRLAB in contrast to the flooded cells. For example, the voltage distribution in gelled VRLAB is larger than in the case of flooded cells.

Many reasons can be attributed for these wide variations. Those due to improper manufacturing practice are:

a. Level of saturation of separators
b. Degree of formation of plates

These can be avoided by accurate filling and proper conditioning cycles.

The electrochemical reason [11] is that the negative plate (anode during discharge, cathode during charge) experiences two competing cathodic reactions, namely, the proton reduction to hydrogen ($2H^+ + 2e^- \rightarrow H_2$) and oxygen reduction to water ($O_2 + 4H^+ 4e^- \rightarrow 2H_2O$). These two reactions have different energy consumption. In the initial stages of lower recombination efficiency, oxygen reduction is accompanied by H_2 evolution at high voltages. Thus, cells with oxygen recombination exhibit decreased voltage (because of the negative plate depolarization, i.e., decrease in potential) while those with lower recombination efficiency exhibit higher voltage. With increasing age, however, the voltage range becomes smaller. Cell voltages may differ between + 300 mV to 130 mV based the median value which is usually 2.25 to 2.3 volts per cell. [201]. However with increasing age, the voltage range becomes smaller [201] Newly installed AGM VR cells may show a variation in individual cell voltages, but once the battery has been in operation on float for about 6 months, the individual cell voltages will stabilize between ±2.5% of the mean applied voltage i.e., between 2.27 and 2.33 VPC, for an applied voltage of 2.30 *VPC* [202].

ACCURATE CONTROL OF CHARGING VOLTAGE

The major complication with the use of VRLAB is that very accurate control must be exercised on the charging system. If the charging voltages rise above a certain limit (>2.4 VPC), the batteries are often unable to recombine all the oxygen evolved. At higher potentials, hydrogen is formed at the negative plate which may lead to pressure build-up resulting in water loss.

Control of the final current during recharge, taking into account the effective recombination capability of the battery, can avoid overcharge without having negative influence on the battery performance. Such controlled recharge method for a cyclically operated battery could be slightly inadequate during the initial stages of the battery life, since, in this case the current that the battery can accept could be higher than the fixed value. For better recharge, in terms of duration and protection of battery, lower maximum currents (less than $0.2C_{20}A$) and lower maximum voltage (less than 2.4 VPC) should be used during the constant current – constant potential period [203]. The lower the recharge current, the more efficient the recharge. Thus only 9.5% capacity can be obtained from a fully discharged battery when it is charged at C/2.5 ampere (i.e., 0.4 CA) till the constant voltage float level is attained, whereas recharge at C/10 (0.1 CA) could give about 80% capacity at the point of reaching the constant voltage float level [79].

The importance of controlling the cell voltages cannot be over-emphasized. Experiments [204] conducted on 48V/100Ah telecommunication VR battery (floated at 2.3 VPC with current of 0.4 – 0.6 mA/Ah (the temperature of end cells, center cell and that of surroundings being equal) indicate:

- That the deliberate short-circuiting of one cell would subject the remainder cells to a float voltage of 2.40 VPC, increases the float current to 2.5 mA/Ah after 24 hours and the temperature by 1°C above the ambient.
- That the deliberate short-circuiting of two cells would subject the remainder cell to 2.51 VPC, increases the current to 11 mA/Ah and the temperature of the battery to 5°C above the surroundings in 24 hours.
- That the deliberate short-circuiting of hree cells would raise the float voltage to 2.63 VPC after 24 hours and the increase in the current would be 50 mA/Ah and rise in the temperature would be 12°C after 24 hours.

- That the deliberate short-circuiting of four cells would raise the float voltage to 2.76 VPC and the current would increase to 180 mA/Ah (in 1 hour) and raise the temperature by 22°C in 1 hour and take the system to thermal runaway condition with concurrent generation of H_2S [204, 205].

However, it is most unlikely that three or four cells would become short circuited simultaneously in a 48 V telecommunication battery. The above data indicate that short-circuiting of 1 or 2 cells would not be disastrous from a thermal point of view.

In a string of several batteries, the variation in voltage will narrow down to ± 2.5% of the mean value within a period of 6 months. When a VRLA cell is overcharged in a CP mode, at different temperatures, the current increases by about 10 folds for a 90 mV increase and by 2 folds for each 10°C rise in temperature.

Provided that VR cells are not used in abusive conditions (e.g., > 60°C and high charging currents or float voltages higher than 2.4 VPC), they do not emit H_2S or SO_2 gas. However, if these conditions are prevailing, these gases can cause contamination and corrosion of copper contacts in electronics equipment [205]. Thus it is essential to monitor the cell voltages of batteries on float.

Impedance measurements well correlated to individual cell design are helpful in obtaining advance warning about unhealthy cells both during manufacturing stages [206] and during operation [207].

Chapter 26

TEMPERATURE/THERMAL RUNAWAY/ DRY OUT

The effect of increased temperature on VRLAB performance is more pronounced than on equivalent flooded electrolyte units. This is due to the inability of the heat to escape from the cells and partly due to the exothermic oxygen-cycle reaction. The AGM VRLABs are more susceptible to thermal runaway phenomenon than the gelled varieties. The battery type, because of the lower interval heat generation (due to lower float currents) and a better heat transfer mechanism, is less susceptible to thermal runaway. The recombination rate of AGM cells is about 20 times higher than that of gelled batteries [11].

However, implementing major improvements to rack design and battery layout will ameliorate this situation. Pressures to comply with environmental legislation that will reduce the use of refrigerant gas in cooling systems will precipitate a move towards fresh air cooling by keeping each cell/battery at least 10-20 mm away from each other and also by other means like use of temperature compensation of flat voltage and by limiting the available current. Matthews [204] has concluded that short circuiting of one or two cells in a string of 24 cells would not be disastrous from a thermal management standpoint. This can be easily detected by regular monitoring of impedance/conductance of the cells. A continuing record of impedance and temperature against time will give an indication of incipient failure.

Drying out is mostly connected with too severe charge regime combined with a high cell temperature. Malfunctioning of the pressure release valve is another reason. “ Dry out” also increases the recombination efficiency thereby increasing the cell temperature.

Forced convection effectively lowers the battery temperature but the temperature difference (ΔT) between internal and skin temperatures of batteries remains substantial in the case of a 48V/100Ah telecommunication battery. Fan cooling for one hour reduced the internal temperature from 45 – 32 °C while the wall temperature fell from 37 - 25 °C [204].

Thompson and Warrel [42] describe experiments comparing the performance of conventional aircraft batteries (employing 6% Sb-Pb grids) and VR aircraft batteries (2V, 25Ah) which had completed 220 cycles. The constant potential charge stability is maintained throughout the life by VR battery, unaffected by age. The increased gassing rate and the possible susceptibility to thermal runaway of ageing conventional battery is well known. The VR aircraft battery shows a low overcharge current and stable temperature of 72°C from half on hour to 3 hours at 14.22V constant potential charge at 50°C, but the temperature of conventional battery begins to show a steep rise right from the beginning of charge and reaches 100°C after 1 hour.

Chapter 27

BATTERY MONITORING

It is needless to emphasize that in order to get a good reliable performance and life, all necessary care has to be taken at the time of installation of the battery itself such as correct selection of charging voltage depending upon the operating temperature and proper inter-cell/module connections. It is necessary to prevent any possible abuse of the battery due to over discharge as well as overcharging excessively. This can be ensured by installing a battery monitoring device. This monitoring device will follow up the battery voltage, temperature, and the charging current. It will effect necessary controls according to the requirements continuously and will cause an alarm signal if warranted. This battery monitoring can be done manually also at regular intervals on selected pilot cells chosen at random and also changing the pilot cell periodically to get an overall picture of the health condition of all the cells/modules in the battery bank. When measurement on the cell/module voltages are made, any abnormal deviation from the normal values that are observed has to be confirmed for any possible defect and the same is to be rectified. Likewise, abnormal temperature rise in any cell/module has to be followed up with corrective measures. The corrective measures for this may be a temporary interruption of charging or lowering the charging voltage. At times, the individual cell/module can be taken out of the bank and replaced by a new one.

Specific gravity of electrolyte cannot be measured in VRLABs. However, conductance measurements can provide some information about the battery capacity degradation [208]. The use of ac impedance measurements at 1 kHz has been well received by battery shop personnel and has been used to complement standard voltage measurements in providing advanced warning of potential problems. However, impedance measurements are sensitive to any change in cell design parameters introduced by manufacturers and hence each battery design must be correlated with impedance measurements.

Monitoring VR automotive lead-acid batteries by sparse-impedance spectroscopy has been reported by Hariprakash [209].

Chapter 28

CATALYST

USE OF RECOMBINING CATALYST PLUGS IN VRLA BATTERIES [210-211] (ELECTROCHEMICAL IMBALANCE)

There is a slow but steady loss of negative plate capacity in valve-regulated (VR) cells. The reasons may be:

The negative in VR cells is more pre-occupied with oxygen-cycle to recombine O_2 to a greater degree than its flooded cell counterparts.

The following reactions are occurring:

(i) O_2 evolution
(ii) O_2 reduction
(iii) Grid corrosion and
(iv) Self discharge.

Electrochemical imbalance results in NAM discharge or "Hydrogen Imbalance" Hydrogen Imbalance is an Electrochemical characteristic of a particular design of valve regulated cells. For "Hydrogen Balance" the rate of positive grid corrosion should be approximately equal to the rate of negative active material self-discharge, which is increased by impurities and temperature and decreased by inhibitors.

During self discharge of NAM, the negative plate releases H_2

$Pb + H_2SO_4 \rightarrow PbSO_4 + H_2$

To nullify the self discharge, the NAM has to be supplied with H_2 (in the form of $H^+ + e^-$) at the same rate as self discharge.

Normal charge Electrons are forced into the negative Active Material.

For NAM "charging hydrogen" ($H^+ + e^-$) should be supplied to nullify the effect of self discharge (due to local action). Usual source of "Charging H_2" is electrolysis (or overcharge)

$$2H_2O \rightarrow 4H^+ + O_2 + 4\ e^-$$

Electrons travel through an electrical circuit. Hydrogen ions migrate into the electrolyte. Oxygen gas travels in the gas space. In a flooded cell, the O_2 exits the cell and the other two components travel to negative active material, the hydrogen leaving the cell after charging the negative active material. In a valve regulated cell, all the three components are recombined and so the negative active material is losing the "charging hydrogen" and hence the negative active material is left uncharged. This will be more so as the cell dries up.

But, how then the negative active material can be kept in a charged condition? The positive grid releases some "charging hydrogen"

$$(Pb + 2H_2O \rightarrow PbO_2 + 4H^+ + 4e^-)$$

(Positive grid corrodes, absorbs O_2 and becomes PbO_2)

The O_2 cycle charges the positive plate without charging the negative plate.

As a lasting solution, William E.M. Jones of Philadelphia Scientific Co. suggests the use of catalyst in each Valve Regulated cell. This is an elegant and simple solution.

The catalyst reacts the H_2 (from self Discharge of Negative active Material) with a proportional amount of O_2 generated at the positive and returns the water to the cell.

Chapter 29

RESULTS OF EXPERIMENTS WITH CATALYSTS

In less than 1 day, the negative plates in a catalyst cell become significantly polarized (charged) and in the non-catalyst cells the negative plates were at or below the open circuit value.

1. The float current of the catalyst cell dropped to less than half of the non-catalyst cells and stabilized there. Gas emissions of catalyst cells were five times lower than the non-catalyst cells.
2. The negative plate gets charged even at as low a voltage as 2.21 volts per cell. While the non-catalyst cells were at open circuit values even at 2.3 volts per cell.
3. Though the catalyst cells emitted lesser volumes of gases, the emissions did not reach zero level. The fundamental reason is positive grid corrosion. Positive grid corrosion consumes oxygen, produces equivalent hydrogen which exits the cell. Therefore in a catalyst cell: Hydrogen emission rate is approximately equal to the positive grid corrosion rate.
4. When the charging was stopped, the gassing rates of the catalyst cells jumped up by a factor of 5 to the same level of non-catalyst cell, while the latter continued to gas uncharged.
5. How to find out if a cell is balanced or not? Unbalanced cells will have reduced gas emission if a "balancing" catalyst is fettled to a mature (RE 100%) valve regulated cell.
6. Approximately 5% of oxygen in the oxygen cycle would be consumed by the catalyst. It will probably be less on a tall cell and more on short ones.

7. If a balanced cell is fitted with a positive grid with lower corrosion rate, the negative plate may discharge and the life will decrease.
8. The more "unbalanced" the cell is, the better is the effect of the catalyst.
9. The theoretical service life charged from under 5 to $>$ 20 years.
10. Earlier failure mode: Negative plate capacity loss.
11. Now with catalyst plug. Positive grid corrosion.
12. Flooded cells: catalyst generates 50W of heat/cell. But VR cells generate $<$ 1W of heat / cell.
13. Lesser chance for catalyst being flooded with water in a valve regulated cell.
14. In a valve-regulated cell, the atmosphere tends to dry the catalyst, should it become wet with condensate.
15. A valve regulated cell with catalyst will exhibit lower voltage.
16. Contaminants unbalance a Valve regulated cell, but a catalyst can make the cell more tolerant of impurities.
17. Small leaks can be taken care of by catalyst. It will not only prevent the negative plate discharge by atmospheric oxygen, but can actually use that oxygen to extend the dry out life of the cell.
18. A failing valve regulated cell in service can be retrofitted with a catalyst to regain the negative plate capacity within a few months.

A novel hydrogen-oxygen recombinant catalyst based on cerium oxide for sealed storage batteries has been investigated by Hariprakash [211]. The role of recombination plugs in VRLA batteries has also been reported by Misra et al. [212]

Chapter 30

FAILURE OF NEGATIVE HALF-BLOCK

There are a number of possible causes of the progressive capacity loss that besets the negative plate of VRLA batteries. These include:

1. Reduction in surface area of the electrodes, possibly following loss of the organic expander, this might be a problem which is peculiar to the VRLA design in that oxygen reaching the negative plate is destructive to organic molecules.
2. Electrolyte stratification, which is likely to be more severe the more open is the pore structure of AGM separator materials.
3. Depression of the potential of the negative plate due to highly efficient oxygen recombination.
4. High hydrogen evolution due to impurities.
5. Recharge inefficiency due to a selective discharge of the plate.
6. In any event, it is encouraging to note that, whichever of these is (are) the culprit(s), the problem of the negative plate is not of universal severity. There are some cases, even in stand-by applications, when a long life is being achieved, and in these instances the degradation of the negative plate is evidently being held at bay.

In most cases the life of the lead-acid battery has been limited by the positive plates [213]. When there were failures of the negative plates, the problems were mainly due to the fact that the pressure release valve was not in good condition. In this case the valve did not close correctly after opening so that oxygen entered the cell and the oxidation of the negative mass gave rise to sulphation during the storage of the battery. To avoid this problem, each pressure release valve has to be fitted to the cell only after carefully controlled testing. This will also avoid the risk due to drying out.

Negative group bar (connector strap) corrosion is a serious problem in some flooded cells [214] and VRLABs [196, 208-209, 218]. A corrosive attack at the lug/connector strap area can be so heavy that the connection is completely destroyed and there is no longer any contact between the plate lug and busbar. This problem can be avoided by proper selection of connector strap alloy and by employing proper technique such as the use of correct flux COS technique. [207, 196, 198, 213, 215-217] The reason for such strap corrosion is that in VRLABs the negative busbar is not cathodically protected during charge sufficiently. [214, 218] The top lead is moistened by very thin and high resistance film creeping up the surface of the plate legs and strops, whereas in flooded cells the strop and lugs of the plate blocks are completely wetted by electrolyte. (However, even in flooded cells, portions of small parts just above the electrolyte level and tiny spots inside the electrolyte were corroded). During charging, the potential becomes equal or close to that of the plates in the negative semi black in flooded cells and so the strap and lugs are protected as a result of reduction of the corrosion product lead sulphate, obtained during battery discharge.

The reaction

$$2\,H^+ + 2e^- \rightarrow H_2$$

occurs on the surface of the negative plates soaked in the electrolyte in addition to the reaction

$$\tfrac{1}{2}\,O_2 + 2e^- + 2\,H^+ \rightarrow H_2O$$

and its rate depends on the solubility of O_2 in sulphuric acid solutions (0.65 mol. dm^{-3} in 5 M solution) and on the diffusion coefficient of oxygen in the sulphuric acid solution ($0.8 \times 10^{-5}\ cm^2\ s^1$ in 5 M solution [219]. Due to its high resistance, the thin liquid film on the connector strap eliminates the effect of polarization on the electrode surface above the AGM level and therefore, the following reactions occur here:

$$Pb \rightarrow Pb^{2+} + 2\,e^- \text{ (discharge)}$$

$$(Pb^{2+} + 2e^-) + \tfrac{1}{2}\,O_2 + H_2\,SO_4 \rightarrow PbSO_4 + H_2\,O \text{ (charge)}$$

$(Pb^{2+} + 2e^-) + H2SO4 \rightarrow PbSO_4 + H_2$ (charge)

The electrolyte at the lug and connector strap becomes diluted. This process of dilution decreases the conductivity of the wetting layer and increases the voltage drop further (i.e., the effect of charging current to convert the $PbSO_4$ is further weakened). Moreover, the pH is increased and as a result of dilution more voluminous corrosion products like tribasic and tetrabasic lead sulphates are formed. However, equalization charges tend to nullify such effects and hence it is imperative to give such periodical equalization charges for float-operated batteries.

The diffusion of oxygen in gaseous phase is faster (0.2 cm^2 s^{-1}) and combined with the abundant availability of oxygen, the reaction occurs at a faster rate. This shifts the mixed (corrosion) potential to more positive values thus accelerating the corrosion. Such a corrosion of a Pb – 2% Sn or Pb – 2% Sn – 0.03% Se alloys can be severe in pure sulphuric acid solution but can be reduced in the latter case by having sodium sulphate in the electrolyte [218].

Experimental recombinant valve-regulated lead-acid batteries that were cycled failed after 250 to 350 deep cycles [220]. The failure was attributed to the negative electrode which showed loss of capacity. When the cells were converted to operation in the flooded mode, they delivered up to 1400 deep cycles. The failure mechanism is assumed to be sulfation due to insufficient recharge. A simple model for the partial currents during recharge is used to analyse the role of recombination and hydrogen and oxygen evolution in the failure mechanism. It is concluded that, at high recombination efficiencies, hydrogen evolution prevents the complete recharge of the negative plate, thus limiting the cycle life.

Chapter 31

PREMATURE CAPACITY LOSS [220-229]

When the VRLA battery was first introduced into deep cycle duty (examples are electic vehicle batteries), its cycle life was very much shorter than that of the flooded battery which it had replaced. In the telecommunications field (infrequent shallow cycling with float charging), too, some experiences with VRLA were disappointing. It became clear that VRLA is not quite a "drop-in replacement" for the flooded battery and probably ought to be treated as a different battery chemistry that happens to offer almost the same voltage characteristics.

The performance defect which was experienced during early deep cycling of VRLA cells was labeled " premature capacity loss", or PCL. It was soon found that more than one mechanism could contribute to PCL.

Premature capacity loss can be categorized into three types: They are described in detail below.

PCL 1: The first form of premature capacity loss to be identified appeared within a few tens of deep-discharge cycles. This occurred with cells containing antimony-free grids and involved the development of a substantial resistance in the vicinity of the interface between the positive grid and the positive active mass. PCL 1 was ameliorated by the incorporation of 1-1.5 wt. % of tin in the alloy used for the positive plate [83, 86].

PCL 2: A second important degradation mechanism involved the swelling of the positive active mass which results in loss of connectivity (hence high resistance) and, ultimately, to shedding. Although this failure mode is well known for flooded batteries, it is aggravated in the case of VRLA batteries for two reasons. In the plane of the plate, the removal of antimony from the positive grid alloy reduced creep strength and allowed the corrosion process to "stretch" the plate. Fortunately, the same addition of tin used to overcome PCL 1 restored the creep strength to the grid alloy. The second potential source of PCL 2 in the VRLA design is the separator. In order to overcome swelling of the active mass in the direction normal to the plane of the plate, it is standard practice to subject the stack of plates to a compressive force maintained by the case walls. The success of this strategy, as a means of preventing swelling, depended on

i. The constraint being transmitted successfully right through the stack of plates; and
ii. The compressive force not being dissipated by collapse of the separator material.

Of the two electrolyte immobilization systems that are prevalent in VRLA batteries, glass mat (AGM) does indeed tend to reduce in thickness when compressed, while gel does not. Gel batteries are characterized by long cycle lives, and novel incompressible separator materials also show promise in this regard.

PCL3: As progress was made towards overcoming the failure mechanisms involving the positive plate, attention was turned to the negative plate. In order to achieve a long effective life for the whole battery, both the plates must retain their full capacity. Of particular concern was the observation that in long deep-cycling tests, the negative plate takes over from the positive as the life limiting element. In particular, it was found that premature failure in float service was often associated with selective discharge of the negative plate – the result of the combined action of a high hydrogen evolution rate and a high oxygen recombination efficiency . A key challenge is the difficulty in bringing the negative plate to a full state of charge in the latter stages of life of a VRLA cell, when the oxygen cycle is becoming progressively more efficient.

Two models have been proposed to explain PCL:

i. Agglomerate of Spheres (Kugelhoufen) Model [222-223]
ii. PAM consists of 3-dimensional network of spherical lead dioxide particles with contact points (called "necks") determining the electronic conductivity. Formation of large necks result in incomplete utilization of PAM. However, treatments like washing, drying and open circuit stand are capable of broadening the neck zones and therefore relax the PCL effect. Hence according to this model PCL is "relaxable insufficient mass use" (RIMU).
iii. Gell crystal concept [224-228]

Lead dioxide particles and agglomerates consist of two zones:

a. Crystal zones which have high electronic conductivity.
b. Amorphous gel zones which have high H^+ or proton conductivity but low electronic conductivity. These zones consist of hydrated linear polymer chains.

These tow zones are in equilibrium with each other. Gel zones are elastic in nature and counteract to suppress the mechanical stress encountered during cycling. Certain additives (e.g. Sb, Co and Sn) increase the electronic conductivity of gel zones.

PASSIVATION OF POSITIVE PLATE (PPP)

This refers to a situation where there is an energetic battier in addition to PCL. This is evidenced by an initial sharp fall in voltage during discharge a rapid rise in voltage during a CC charge and a very low current during CP charging.

Rand [229] differentiates four types of PPP. They are:

1. Thermopassivation which is caused by exposure to excessive drying temperature for long duration. To avoid such a type of PPP, the plates should be dried between 55 and 65°C and the plates should not be bone dried (the moisture content: 0.25 – 0.4%).
2. Storage passivation is caused by storing batteries for extended periods without freshening charge.
3. Float passivation is caused by continuous float operation without periodic discharge, and
4. Squeeze passivation occurring due to deep discharge of very low currents.

LEADING THROUGH

"Leading through" occurred in certain types of VRLAB which experienced frequent deep discharges. Because of the increased solubility of the lead sulphate in dilute electrolytes, the pores become clogged with lead fibres which short-circuit the plate blocks. Although "leading" through separators was a reason for failure of earlier generation of VRLABs, this has been eliminated by incorporation of sulphate ions in the electrolyte.

If processing is not carried out properly during manufacturing stage, leading through will happen regardless of the tortuosity of the separator. The mechanism of leading through is that there are areas of relatively high pH or low acidity when batteries are poorly filled. In these areas lead oxide or lead sulphate will go into solution. During the formation or recharge process, dendrites are formed.

WATER LOSS

Water loss from VRLAB occurs by three routes:

1. recombination inefficiency
2. permeation through the container material
3. grid corrosion

Under float conditions, recombination efficiencies are very high and water loss by this route is negligible. But in cycling service, considerable gassing occurs with each charge. This water loss can be lessened by keeping the potential at a low valve, but the charge duration is then long and inefficient. Higher voltage enables rapid charging but causes significant water loss. A charger which automatically reduces the charging voltage after a given time would be helpful in overcoming this problem.[230]

In automotive batteries, even a small loss of water will cause a large drop in battery voltage at high currents. This will reduce cranking performance and eventually results in the failure of the battery. In low rate application water loss can be tolerated to a greater extent but the resulting higher acid concentration increases the corrosion rate and battery failure occurs as a result of the combination of both effects.

POSITIVE MASS SOFTENING AND SULPHATION

Batteries subjected to cycling application face the problem of sulphation and positive mass softening. The two can be distinguished from the measurement of OCV after a full charge. Softening does not produce any reduction in acid density in the cell. During cycling PbO_2 particles lose contact with each other, so that parts of the active mass become electrically isolated and cannot take part in the discharge process. The available capacity is less. Softening is promoted by high discharge rates, but sometimes small discharge rates also causes softening. Good process control during manufacturing is essential to void softening.

Short Circuits [69, 208]

Short circuits inside the battery results in the battery failure. Short circuits occur through the separator by lead needles formed when the electrolyte gravity reaches very low levels as a result of over-discharge. Additives such as sulphate salts to the electrolyte overcome this problem to a great extent, particularly sodium sulphate prevents the hydration shorts that is likely to occur in the electrolyte filling process.

In valve regulated batteries, there is no mud space and plates rest directly on the bottom of the case. When glass mat separator is used as a single sheet, short circuits are likely to occur. Enveloping of either the positive or negative plates by the separator will help in overcoming this problem.

To increase reliability and decrease the problem of battery failure during its expected life time, the batteries require a manufacturing process that has to be more precisely controlled than that employed for flooded batteries. Minor mistakes can have tremendous consequences for VRLAB. Many failures are caused by the incorrect use of the battery. So the users are to be well aware of the procedures for the care and maintenance of the battery.

Methods of absorbing the longitudinal growth of tall plates using stub-foot [231] and a suspension system for plates [80]. In the latter work, a sturdy injection-moulded PP component fits on the element via lugs on the grids. There is pocket moulded into the part that coincides with positive, and this allows for growth. In the stub-foot design, the feet of the positive plates are holed to relieve the longitudinal stress during vertical growth.

Analysis of Failure Modes

Failure modes have been analyzed by Ball and his co-workers. [232] They concluded that at failure there is a variation in the capacity of each cell and it was often possible to pinpoint a good and bad cell within a battery. The electrode potentials of the good and bad cells indicated that reduction in capacity could usually be attributed changes in the positive as opposed to the negative electrode. Examination of polished cross-section of positive grid bars showed an increase in corrosion layer thickness on the sides of the bar. Finite element analysis revealed that this area corresponds closely to the region of highest current density. The corrosion layer thickness was shown consistently to be greater on the positive electrodes in the bad cells when compared to the good cells. Detailed examination of the corrosion layers indicated a number of characteristic defect types. This suggests that more than one mechanism is contributing to the corrosion process. Oxygen concentration within the corrosion layer was shown to decrease closer to the grid wire. This lack of stoichiometry correlates with an increase in resistance.

CONCLUSION

The use of VR batteries is increasing at a tremendous rate in all the application areas. Developments are taking place day by day. An attempt has been made in the above pages to reveal to the readers the manufacturing procedures and the developments that have so far taken place. As pointed out earlier, the developmental process is a continuing one and it is expected that the VR batteries will occupy the foremost place in the realm of application of lead-acid batteries.

The manufacture of VRLAB requires a more precisely controlled processing and special materials than those employed for flooded batteries, which mean a higher production cost. Hence, the initial investment on VRLAB is more. However, due to the low maintenance-cost, the cost ultimately works out in favour of VRLAB.

REFERENCES

[1] Gaston Planté, Nouvelle pile secondaire d'une grande puissance, *Compt Rend* 1860, 50, 640 ; Recherches zur l'élecricité, Gouthier-Villars, Paris, 1883, Ch. II, p. 29.

[2] Volta, A, *Phil. Trans. Roy. Soc.*, 1800, 90, 403.

[3] B.D. McNicol, B.D; Rand, D.A.J. in McNicol, B.D; Rand, D.A.J. (eds.) *Power Sources for Electric Vehicles*, Elsevier, Amsterdam, 1984, p. 56.

[4] Clive D.S. Tuck Ed, *Modern Battery Technology,* Ellis Horwood, New York, 1991, p. 1.

[5] Faure's Secondary Battery, *Electrician*, 1881, 6, 323; *French Patents* 1880, 139 258; 1881, 141, 057; *German patent* 1881, 10 926 ; *British Patent* 1881, 129; *U.S. Patent* 1882, 252 002 (1882).

[6] Sellon, *British Patent* 1881, 3987.

[7] Koredsch, K.V. (Editor) *Batteries,* Vol.2, *Lead Acid Batteries and Electric Vehicles,* Marcel Dekker, NY1977, 212.

[8] Glasstone, J.H; Tribe, A, *Nature*, 1881, 25, 221 and 461.

[9] Glasstone, J.H; Tribe, A *Nature*, 1882, 26, 251, 342 and 602.

[10] Vinal. G.W, *Storage Batteries*, John Wiley, 1954.

[11] Hans Tuphorn. *J. Power Sources*, 1993, 46, 366.

[12] Bearinger, I.C. *The Battery Man*, 1992, Dec, 28-35.

[13] Baars, E.G. *Proc. 12th Power Sources Conf.*, 1958, U.S. Army Signal Res. And Dev. Lab New Jeresy.

[14] Dehmelt, K. *Proc. 13th Power Source Conf.*, 1959, U.S. Army Signal Res. And Dev. Lab., New Jeresy.

[15] Walinder, J.A."*Lead-Acid Batteries. The Use of Catalators*", Lead Development Association, London, 1965, pp 12-16.

[16] Watanabe, M; Yonesu, K. Development and Present Status of the Catalyst Plug for the Stationary Lead-Acid Battery in Japan, *Technical Leaflet*, 1978, Japan Storage Battery Co., Kyoto, Japan.

[17] Ruetschi, P *US Pat.* 1960, 2151106.

[18] *Japanese Pat.* 1968, 5291283.

[19] P. Ruetschi, P; Ockerman, J.P. *J. Electrochem. Soc.* 1969, 116, 1222.

[20] *Japanese Pat.* 1969, 536923.

[21] Lange, A.E. *US Pat.* 1938, 2 131 552.

[22] Levin, A; Thompson, W.S. *Brit. Pat.*1943, 571820.

[23] Jeannin, R. *French Pat.* 1950, 102 709.

[24] Neumann, G; Gottesmann, U. *US Pat.* 1951, 2, 571, 927.

[25] Bureau Technique Gautrat *Brit. Pat.* 1952, 677 770.

[26] Jeannin, R. *U.S Pat.* 1953, 264455, 17.

[27] Devitt; McClelland, *US Pat.*1975, 3862861.

[28] Kobayashi, Y; Sakata, Y; Fukuda, S. *Progress Batteries and Solar Cells*, 1987, 6, 133-135.

[29] Hans Tuphorn, *Progress Batteries and Solar Cells*, 1987, 6, 145-149.

[30] Kotera, T; Nambu, A. *Progress Batteries and Solar Cells*, 1987, 6, 119-124.

[31] Kiessling, R. *J. Power Sources*, 1988, 23, 135-141.

[32] Sato, H. *J. Power Sources*, 1989, 28, 173-180.

[33] Yamada, T; Nakazawa, Y; Tsujino, N. *J. Power Sources*, 1992, 38, 123-136.

[34] Hans Tuphorn, *J. Power Sources*, 1992, 40, 47-61.

[35] Arias, J.L; Rowlette, J.J; Drake, E.D. *J. Power Sources*, 1992, 40, 63-72.

[36] Rowlete.J.J; Harbaugh, D.L. *Proc. 7th Annual Battery Conf. Applications and Advances*, Jan 1992, California State Univ., Long Beach, CA paper 92 EV-4.

[37] Masaki Shiomi; Katsuhiro Takahashi; Masahara Tsubota, *J. Power Sources*, 1993, 42, 173-184.

[38] Calasanzio, D; Cecchinoto, G; Marchetto, M. *J. Power Sources*, 1993, 42, 247-257.

[39] Stevenson, J.M; Wilson, F; Dyson, J.I. *J. Power Sources*, 1991, 33, 135 – 143.

[40] Sunu, W; Burrows, B.W. in Thompson, J (ed.), *Power Sources 8*, Academic Press, London 1981, p 601.

[41] Szymborski. J; Burrows, B.W in Thompson, J (ed.), *Power Sources 9*, Academic Press, London 1983, p 113.

[42] Thompson,J; Suzanne Warrel, in Szymborski, J; Burrows, B.W. in Thompson, J (ed.), *Power Sources 9*, Academic Press, London 1983, p 97.

[43] Wagner, R. *J. Power Sources*, 1997, 67, 163-172.

[44] Narasimhan, L; Raj, P; Hussain, Z. *J. Power Sources*, 1999, 78, 214-219.

[45] Nakayama,Y; Takahashi, S; Hirakawa, K. Yamaguchi, Y. *J. Power Sources*, 2004, 125,135-140.

[46] Bhardwaj, R.C; Than, J. *J. Power Sources*, 2000, 91, 51-61.

[47] Chen, J.S. *J. Power Sources*, 2000, 85, 172-177.

[48] Nakayama, Y; Hojo, E; Koike, T. *J. Power Sources*, 2003, 124, 551-558.

[49] Caballero, A.; Cruz, M; Hernan, L; Morales, J; Sanchez, L. *J. Power Sources*, 2004, 125, 246-255.

[50] Wade, E.J., *Secondary Batteries,* Electrician Printing and Publishing Company, London, 1902.

[51] Vinal. G.W, *Storage Batteries*, John Wiley, 1954.

[52] Bode, Hans, *Lead-Acid Batteries,* John Wiley, New York, 1977.

[53] Crompton, T.R., *Small Batteries*, Vol.1, John Wiley and Sons, New York, 1982.

[54] Pavlov, D in McNicol B.D; Rand, D.A.J (eds.) *Power Sources for Electric Vehicles*, Elsevier, Amsterdam, 1984.

[55] Barak, M. *Electrochemical Power Sources*, *IEE Series*, Peter Peregrinus Ltd., stevenage, London, 1984.

[56] Hehner, N.E. 3rd Edition revised by Orsino, J.A. Storage *Battery manufacturing Manual III,* IBMA, Largo,Florida, USA 1986.

[57] Keihene, H.A., *Battery Technology Handbook,* 1989.

[58] Peters, K. in Clive D.S. Tuck (ed.) *Modern Battery Technology* Ch. 4, 1991.

[59] Crompton, T.R., *Battery Reference Book*, Second Edition, SAE International, P.A., USA,. 1996.

[60] Rand, D.A.J; Woods, R; Dell, R.M. *Batteries for Electric Vehicles,* Research Studies Press Limited, Taunton, Somerset, England, John Wiley and Sons Inc., New York, 1998.

[61] Salkind, A.J.; Anthony G. Cannone; Forrest A. Trumbore. in *Handbook of Batteries*, third edition (Ed.) David Linden and Thomas B. Reddy, McGraw-Hill, New York, 2002, pp. 23.1-23.88.

[62] Dietrich Berndt, "*Maintenance-Free Batteries*" Research Studies Press Limited, Taunton, Somerset, England, John-Wiley and Sons Inc., New York 1993 (First Edition), 1997 (Second Edition).

[63] Salkind, A.J; Ronald O. Hammel; Anthony G. Cannone; Forrest A. Trumbore. in *Handbook of Batteries*, third edition (Ed.) David Linden and Thomas B. Reddy, McGraw-Hill, New York,2002, pp. 23.1 -23.46.

[64] Rand, D.A.J; Moseley, P.T; Garche. J ; Parker, C.D.(Eds.) *Valve-Regulated Lead –Acid Batteries,* Elsevier, New York, 2004.

[65] Milner, P; Thomas, U.B. in *Advances in Electrochemistry and Electrochemical Engg*. (ed) Tobias, C.W. Vol. 5, p 69, 1967.

[66] Mahato, B.K; Laird, A.C. *J. Electrochem Soc*. 1974, 12, 113.

[67] Atkin, J; Bonnaterre, R.; Laurant, J.F. "*Power Sources 6*" (ed,) D.H. Collins, Academic Press, London 1976, p 91.

[68] Bullock, K.R in Bullock, K.R; Pavlov, D. (eds.) *Advances in Lead Acid Batteries, Proc*. Vol.84-14. *The Electrochem. Soc*. Inc., New Jeresy, 198.

[69] Hardman, A.M. *J. Power Sources*, 1988, 23, 127-134.

[70] Varma, B.P; Boden, D.P. *The Battery Man*, 1986, Dec, p 12-18.

[71] Eberhard Nann, *J. Power Sources*, 1991, 33, 93-103.

[72] Hans Tuphorn, *J. Power Sources*, 1992, 40, 50.

[73] Berndt, D. *J. Power Sources*, 2001, 95, 2-12.

[74] Nelson, R., *J. Minerals, Metals and Materials Soc*., 2001, 53, 28-33.

[75] Balakrishnan, P.G. *Proc. First National Sem. Maintenance-free Lead-Acid Battery and Applications,* SAEST and CECRI, 15-16 Feb 1994, Chennai.

[76] Rand, D.A.J; Woods, R; Dell, R.M. (Eds) *Batteries for Electric Vehicles,* Research Studies Press Limited, Taunton, Somerset, England, John Wiley and Sons Inc., New York, 1998, p. 157.

[77] 76. Dietrich Berndt, "*Maintenance-Free Batteries*" Second Edition, Research Studies Press Limited, Taunton, Somerset, England, John-Wiley and Sons Inc., New York, 1997, p. 98, 262.

[78] Peters, in Manders, J.E.; Lam, L.T; Peters, K; Prengaman, R.D; Valeriote, E.M., *J. Power Sources*, 1996, 59, 205 .

[79] Weighall, M. J. *J. Power Sources*, 2001, 95, 209-217.

[80] Don Hosking, *J. Power Sources*, 1993, 45, 111-117.

[81] Holden, L.S. *J. Power Sources*, 1996, 59, 192.

[82] Li-ZH; Guo-YL; Wu-LZ; Perrin-M; Doering-HL; Garche-J. *J. Electrochemical Soc.,* 2002, 149, A934-A938.

[83] Dietrich Berndt, "*Maintenance-Free Batteries*" Second Edition, Research Studies Press Limited, Taunton, Somerset, England, John-Wiley and Sons Inc., New York, 1997, p. 57.

[84] Albert, L; Chabrol, A; Toreheux, L; Steyer, PH; Hilger, J.P. *J. Power Sources*, 1997, 67, 257-265.

[85] Moseley Patrick, T. *J. Power Sources*, 1998, 73, 122-126.

[86] Smith, K.M., Morgan K.A., in *Proc. 13th Annual Battery Conference on Application and Advances,* January 13-16, 1998, Long Beach, CA, 179-185.

[87] Rocca, E; Steinmetz, J; Weber, S. *J. Electochem Soc*. 1999,146, 54-58.

[88] Patrick T. Moseley, *J. Power Sources*, 1999, 80, 1-6.

[89] Chang,TG; Jochim, DM. *J. Power Sources*. 2000, 91, 177-192.

[90] Giess, H. *J. Power Sources*, 1995, 53, 31-43.

[91] Prengaman, R.D. *J. Power Sources*, 1995, 53, 207-214.

[92] Takahasi, K; Yasuda, H; Hasegawa, T; Horie, S. Denki Kagaku, 1998, 66, 308- 313.

[93] Shimoura, I; Terada, M; Yamaguchi, Y; Takashima, M. *Electrochemistry* 2000, 68, 967-971.

[94] Torcheux, L; Villaron, A; Bellmunt, M; Lailler, P. *J. Power Sources*. 2000, 85, 157-163.

[95] Taguchi, M; Hiramatsu, Y; Hirasawa, T; Tanaka, N. *J. Japan Instt Metals*, 2001, 65, 720-725.

[96] 95. Ball, RJ; Kurian, R; Evans, R; Stevens, R. *J. Power Sources*, 2002, 111, 23-38.

[97] Ball, RJ; Kurian, R; Evans, R; Stevens, R. *J. Power Sources*. 2002, 103, 207- 212.

[98] Ball, RJ; Kurian, R; Evans, R; Stevens, R. *J. Power Sources* 2002, 109, 189-202.

[99] Wang, J; Guo, ZP; Zhong, S; Liu, HK; Dou, SX. *J. Applied Electrochem.* 2003, 33, 1057-1061.

[100] Lakshmi, C.S, *J. Power Sources*, 2000, 88, 19-26.

[101] Prengaman, D. in Bullock, K.R; Pavlov, D (eds)., *Advances in Lead Acid Batteries, Proc.* Vol. 84-14, The Electrochem. Soc. New Jeresy, 1984.

[102] Archer, *The Battery Man*, 1983 October p 23.

[103] Edward J. Schmidi, *The Battery Man*, 1985, March p 15.

[104] Bolotovski, V.I., *J. Power Sources*. 1991, 33, 51.

[105] Gillian, W.F. *J. Power Sources* 1987, 19, 143.

[106] Cui Ronglong, *J. Power Sources* 46 (1993) 327-333.

[107] Valeriote, E. M., in "Lead-Acid Battery Technology" in Manders, J.E.; Lam, L.T; Peters, K; Prengaman, R.D; Valeriote, E.M., *J. Power Sources*, 59 (1996) 200-201.

[108] Rodyakin, V.V; Sventne, t. *Metally*, 1958,31,43.

[109] Maslanka-Ormanowa, *Z. Prace. Inst. Minist. Hutn.* 1953, 5, 89.

[110] Dandapani, K.S; Renganathan, N.G; Kamaludeen, M; Augustin, C.O; Desigan, *P.S. Bull. Electrochem.*, 1990, 6, 457.

[111] Prenagmann, R.D. *U.S. Patent* 1984, 4 439 398.

[112] Prenagmann, R.D. in Bullock, K.R; Pavlov, D (Eds), I *Advances in Lead-Acid Batteries, Proc.* Vol. 84-14, The Electrochem. Soc., NJ, 1984, p. 205.

[113] J.T. Nawracay, *The Battery Man*. 1988, Nov. pp 60-66.

[114] J.T. Nawracay, *J. Power Sources*. 1988, 23, 73-77.

[115] Hibbins, S. H; Closset, B; Bray, M. *J. Power Sources* 1995, 53, 78-82.

[116] Zhong-S; Liu-HK; Dou-SX ; Skyllaskazacos-M. *J. Power Sources*, 1996, 59, 123-129.

[117] K. Takahasi, H. Yasuda and T. Hasegawa, *Prog. Batteries and Solar Cells*, 1987, 6,126-129.

[118] Ball, R.J; . Evans, R; Stevens, R. *J. Power Sources*. 2002, 103, 213-222.

[119] David Prengaman, R. *J. Power Sources*, 1995, 53, 207-214.

[120] Bouton, G.M; Phipps, G.S. *Proc. 92nd General Meeting of the Electrochemical Soc.*, Boston, October 1947, 45.

[121] "The St. Joe Ternary Bloom Test", *Technical Leaflet*, St. Joe Minerals Corporation, New York City, USA, 1976.

[122] N.E. Hehner, 3rd Edition revised by J.A. Orsino, *Storage Battery Manufacturing Manual* III, IBMA, Largo,Florida, USA 1986.

[123] Clarke, B.L; Wooten, .L.A. *Ind. and Chem.*, Analytical Edition, 1933, 5, 313.

[124] Dyson, I; Griffin, P. J. *J. Power Sources*, 2003, 116, 263-282.

[125] Lam, L.T; Lim, O.V; Haigh, N,P; Rand, D.A.J; Manders, J.E; Rice, D.M, *J. Power Sources* 1998, 73, 36-46.

[126] Yeh,C.H; Wan, C.C; Chen, J.S. *J. Power Sources*, 2001, 101, 219-225.

[127] Wang-J; Zhong-S; Wang-XL; Liu-HK; Dou-S. *J. Power Sources*, 2003, 122,195- 200.

[128] Wang, J; Zhong, S; Wang, G.X; Bradhurst, D.H; Ionescu, M; Liu, H.K; Dou, S.X; *J. Alloys and Compounds* 2001, 327, 141-145.

[129] Ball-RJ; Evans-R; Thacker-EL; Stevens-R. *J. Material Science*, 2003, 38, 3013-3017.

[130] Lam,L.T; Haigh, N.P; Lim, O.V; Rand, D.A.J; Manders, J.E, *J. Power Sources*, 1999, 78, 139-146.

[131] Lam,L.T; Haigh, N.P; Rand, D.A.J; Manders, J.E, *J. Power Sources*. 2000, 88, 2-10.

[132] Lam, L.T; Haigh, N.P; Rand, D.A.J., *J. Power Sources* 2000, 88, 11-17.

[133] Satevenson, M.W; Manders, J.E; Eckfield, S; Prengaman, R.; *J. Power Sources*, 2002, 107, 146-154.

[134] Boden, D.P; Arias, J; Fleming, F.A; *J. Power Sources*, 2001, 95, 277-292.

[135] Manders, J.E; Lam, L.T; Peters, K; Prengaman, R.D; Valeriote, E.M. *J. Power Sources* 1996, 59, 203.
[136] Lam, L.Y; Rand. D.A.J, *Batteries Int* 1992, 13, 52-57.
[137] Rand. D.A.J; Lam, L.Y, *The Battery Man.* 1992, Nov. 28-28.
[138] Aidman, G.I. *J. Power Sources*, 1996, 59, 25-30.
[139] Devitt, J. *J. Power Sources*, 1997, 64, 153-156.
[140] Culpin, B., *Journal Power Sources* 1995, 53, 127-135.
[141] Zguris, G.C., *Journal Power Sources* 2000, 88, 36-43.
[142] Ball, R.J; Kurian, R; Evans, R; Stevens,.R. *Journal Power Sources*. 2002, 104, 234-240.
[143] Weighall. M.J., *J. Power Sources*, 2003, 116, 219-231.
[144] Badger, J; Wandry, K; C. Ind, C. *Prog. Batteries and Solar Cells*, 1990, 9, 44-47.
[145] Advt. Inserted by Hollingsworth and Vose Co., USA, *The Battery Man*, 1991, Nov. p 27.
[146] Zguris, G.C., *Batteries Int.* 1991, 7, 80-81.
[147] Zguris, G.C. *The Battery Man*, 1992, Sept. 42-48.
[148] Matthes, K; Keily, K; Baxtor, B.W.(ed.) *Power Sources 12,* Leatherhead, England, 1988.
[149] Kevin Campbell, *Advances in Production Charging : Single Battery* Formation, Bitrode Inc., 1998.
[150] Lawrence, E.L; Ball, R.J; Evans, R; Stevens, R. *J. Power Sources* 2002, 110 125-132.
[151] Barak, M. *Electrochemical Power Sources, IEE Series*, Peter Peregrinus Ltd., Stevenage, London, 1984 p 302.
[152] Yuasa Booklet on *NP-Series MP Sealed Lead Acid Batteries.*
[153] Thompson, J; Suzanne Warrel. in Thompson, J Ed. *PowerSources 9,* Academic Press, 1983, p.97.
[154] Ruetschi, P. *J. Power Sources* 1977/78, 2, 3-24.
[155] Eberts, K. in DH Collins (ed) *Power Sources 2*, Pergaman Press, London, 1970, p 69.
[156] Ruetschi, P. *US Pat.* E, 124, 488 (March 10, 1964).
[157] 156. Holden, L.S; in Rand, D.A.J; Holden, L.S; May, J.G; Newnham, R.H; Peters. K. *J. Power Sources*, 1996, 59, 191-197.
[158] 157. Oh, S.H.; Kim, M; Lee, J.B; Lee, H.L; *Bull. Korean Chem., Soc.* 2002, 23, 75-80.
[159] Martha, S.K; Hariprakash, B.; Gaffoor, S.A; Shukla, A.K; *Bull. Material Science*. 2003, 26, 465-469.

[160] Schwetz, A. *Proc. LABAT -93*, 7-11 June 1993, St. Konstantine, Varna, Bulgaria, Abstract No. 43.
[161] Crouch, D.A; Reitz, J.W. *J. Power Sources*. 1990, 31, 125-133.
[162] Wandzy, K.J.; Taylor, G.W. *The Battery Man*. 1986 July p 14-18.
[163] J. Badgor, *IBMA 50th Convention*, Chicago, Il., USA October 27-29, 1987.
[164] Peters, Ken. *J. Power Sources* 1993, 42, 155-164.
[165] *Technical data leaflet* Hollingsworth and Vose Co., USA. 1992.
[166] *Battery International*, July 1992.
[167] Holden, L.S., in Rand, D.A.J., Holden, L.S., May, J.G., Newnham, R.H., and Peters. K, *J. Power Sources*, 1996, 59, 191-197.
[168] Hans Tuphorn, *J. Power Sources* 1988, 23, 143-145.
[169] Kugel, M. *Ger. PAT* 1926, 480149.
[170] Voss, E. *J. Power Sources*, 1988, 24, 171-184.
[171] Ferreira-AL; Lingscheidt-HA. *J. Power Sources*, 1997, 67, 291-297.
[172] Vinod-MP; Vijayamohanan-K; Joshi-SN. *J. Power Sources*, 1998, 70, 103-105.
[173] Oh-SH; Kim-M; Lee-JB; Lee-HL. *Bull. Korean Chem. Soc.* 2002, 23, 75-80.
[174] K. Peters, in Clive D.S. Tuck (ed.) *Modern Battery Technology* Ch. 4, p. 236.
[175] Paul Conrad, *Battery Materials*, Noyes Data Corporation, NJ, USA, 1970, pp. 121-127 (*US Pat*ent 3 305 396, 1967).
[176] Dietrich Berndt, "*Maintenance-Free Batteries*" Second Ed., Research Studies Press Limited, Taunton, Somerset, England, John-Wiley and Sons Inc., New York, 1997, P. 317.
[177] John A. Mills, *J. Power Sources*, 1992, 38, 137-141.
[178] The ALABC, *Technical Information* (1994).
[179] R.F. Nelson, *The Battery Man*, 1993.
[180] Pavlov, D; Bashtavilova, E. *J. Electrochem. Soc.*, 1986, 133, 241-248.
[181] K. Takahashi, M. Tsuboto, K. Yonezu and K. Ando, *J. Electrochem. Soc.*, 1983, 130, 2144-2149.
[182] Hollenkamp, A.F. *J. Power Sources*, 1999, 59, 87.
[183] Mcgregor, K; Hollenkamp, A.F; Barber, M.; Huynh, T.D; Ozgun, H; Phyland, C.G; Urban, A.J; Vella, D.G; Vu, L.H; ALABC Project No. AMC-009, Internal Report, 1997. Also see *J. Power Sources*, 1998, 73, 65-73.
[184] T.G. Chang, E.M. Valerioto and D.M. Jochim, *J. Power Sources*, 48 (1994) 163.

[185] K. Tomantschger, E.V. Valeriote, J.S. Klarchuk, T.G. Chang, M.J. Dewar, V. Ferrone and D.M. Jochim, *Proc. 13th Annual Battery Conference on Application and Advances*, January 13-16, 1998, Long Beach, CA, 173-178.

[186] J.K. Nor and J.L. Vogt, *Proc. 13th Annual Battery Conference on Application and Advances*, January 13-16, 1998, Long Beach, CA, 191-197.

[187] Chang-TG Jochim-DM. *J. Power Sources*, 2000, 91, 177-192.

[188] Soria, M.L; Valenciano, J; Ojeda, A; Raybaut, G; Ihmels, K; Deiters, J; Clement, N; Morales, J; Sanchez, L. *J. Power Sources*, 2003,116, 61-72.

[189] Moseley, Patrick T., *J. Power Sources*, 2004, 133, 104-109.

[190] Cooper, A. *J. Power Sources*, 2004, 133, 104-109.

[191] Wagner, R. J; Sauer, D.U. *J. Power Sources*, 2001, 95, 141-152.

[192] S. A.Shivashankar, A. K. Shukla, A.U. Mane, B. Hariprakash and S. A. Gaffoor, *US Patent Application,* January 31, 2003.

[193] B. Hariprakash, A. U. Manc, S. K. Martha, S. A; Shivashankar, S. A. Gaffoor and A. K. Shukla, *Electrochem. Solid-State Lett.*, 7 (2004) A66.

[194] S. K. Martha, B. Hariprakash, S. A. Gaffoor and A. K. Shukla, *9ELBC* (2004).

[195] S. K. Martha, B. Hariprakash, A. K. Shukla and S. A. Gaffoor, *Indian Patent* (Under process).

[196] Brecht, W.H; O'Leary, *N.F. INTELEC, 1998 Conf.* San Diego, CA, USA, Oct 1988.

[197] May, G.*J. Power Sources*, 1995, 53, 111-117.

[198] Lailler, P; Zaninotto, F; Nivet, S; Torcheux, L; Sarrau, J.F; Vaurijoux, J.P; Devilliers, D. *J. Power Sources* 1999, 78, 204-213.

[199] Hebber, R; *Proc. 1st MF Lead-Acid Battery and Applications*, SAEST-CECRI, Feb.15-16,1994, Madras, pp .5-8].

[200] Yuasa booklet on *VR batteries* NP Series., 1999.

[201] Bernardi-DM Ying-RY Watson-P. *J. Power Sources*, 2004,151, A85-A100.

[202] Handling instructions for stationary Lead acid batteries Sonnenschein Accumulatoren Manchester, England, *Publication No.TA1,* Dea.1991. Model Dryfit A600 and Monobloc batteries model dryfit Block.

[203] Specifiers manual, *Chloride Power Safe*, Chloride Industrial Battery Group Publication SP 51/5M/269 Nov.1990].

[204] Calasanzio-D Caselli-M Ghiotto-D *J. Power Sources*, 53(1995)143-147.

[205] Matthews, K; Papp, B; Nelson, R.F. in Keily, T; Baxter, B.W.(eds) "*Power Sources 12*" International Power Sources Symp. Committee, Leatherhead, England, 1989, pp1-31.

[206] Robinson, R.S; Tarascon, J.M. *J. Power Sources*, 1994, 48, 277-284.

[207] Verville, G; Roberge, P; Smit, J. in *Power Sources 12*, International Power Sources Symp. Committee, Leatherhead, England, 1989, pp-59.

[208] Kniveton, M.W. *J. Power Sources*, 1995, 53, 149-152.

[209] Michael Cox, Midtronics, Inc. *Electrical Construction and Maintenance*, Feb 1,2002.

[210] B. Hariprakash, S. K. Martha and A. K. Shukla, *Proc. Indian Acad. Sci.* (*Chem. Sci.*), 115 (2003) 465.

[211] *Battereis Int.* April 1999.

[212] B. Hariprakash, P. Bera, K.C. Patil, M.S. Hegde. A.K.Shukla and S.A. Gaffoor, *Indian Patent Application,* April 24, 2000.

[213] Misra-SS Noveske-TM Mraz-SL Williamson-AJ. *J. Power Sources* 2001, 95, 162-173.

[214] Rainer Wagner, *Proc. 11th International Lead Conference*, Venice, Italy, 1993.

[215] Omae, T; Osumi, S; Takahashi, K; Tsuboto, M. *J. Power Sources*, 1997, 65, 65-70.

[216] Baudo, G; Cecchinnato, G.B. *Proc.11th INTELECT Meeting*, Florence, IEEE Society, 1989, Paper12.3.

[217] Prengaman, R.D. *4th International Lead Acid Battery Seminar*, San Francisco, ILZRO, 1990, p.19.

[218] Dietrich Berndt, *Maintenance Free Batteries*, Research Studies Press Ltd., Tauton, Somerset, England, 1993, p.62-63.

[219] Pavlov, D; Dimitrov, D; Petkova, G; Gnehm, H.G.C. *J. Electrochem. Soc.* 1995, 142, 2919.

[220] Mrha, J., Micka, J., Jindra, J., and Musilova, M., *J. Power Sources* 1989, 27, 91.

[221] Atlung, S; Zachauchristiansen, B. *J. Power Sources*, 1994, 52, 201-209.

[222] Mosley, P.T. *J. Power Sources*, 2001, 95, 218-223.

[223] Winsel, A; Voss, E; Hullmeine, U. *J. Power Sources*, 1990, 30, 209-226.

[224] Bashtabelova, E; Winsel, A. *J. Power Sources*, 1995, 53, 175-183.

[225] Hollenkamp, A.F. *J. Power Sources*, 1991, 36, 567-585.

[226] Pavlov, D. *J. Power Sources*, 1995, 53, 9-21.

[227] Pavlov, D; Balkanov, I. *J. Electrochem. Soc.* 1992, 139, 1830-1835.

[228] Pavlov, D. *J. Electrochem. Soc.* 1992, 139, 3075-3080.

[229] Pavlov, D. *J. Power Sources*, 1993, 42, 345-363.

[230] Rand, D.A.J; Hollenkamp,A.F; Nelson, R.F, *The Battery Man*, Sept. 1993, p.16.
[231] Takahashi, K; Hohihara, N; Yasuda, H; Ishi, T; Jimbo, *J. Power Sources*, 1990, 30, 23-31.
[232] Iwata, M. *Progress Batteries Solar Cells*, 1989, 8, 231-234.
[233] R.J. Ball and R. Stevens, *J. Power Sources*, 2003, 113, 228-232.

INDEX

A

B

C

D

E

F

G

H

I

U

V

W

X

Y

Z